Eradication
of Exotic Pests

Eradication of Exotic Pests

Analysis with Case Histories

Donald L. Dahlsten and Richard Garcia, Editors

Hilary Lorraine, Associate Editor

Yale University Press

New Haven and London

Published with assistance from the Kingsley Trust
Association Publication Fund established by the Scroll
and Key Society of Yale College.

Designed by Jo Aerne and set in Galliard type by
The Composing Room of Michigan, Inc.
Printed in the United States of America by
Thomson-Shore, Inc., Dexter, Michigan.

Library of Congress Cataloging-in-Publication Data
Eradication of exotic pests: analysis with case histories
    Donald L. Dahlsten and Richard Garcia, editors;
    Hilary Lorraine, associate editor.
        p.   cm.
    Includes bibliographies and index.
    ISBN 0-300-04332-5
    1. Pests—Control.   2. Pests—Control—Case
studies.   I. Dahlsten, Donald L., 1933– .   II. Garcia,
Richard, 1930– .
SB950.E78   1989   88-37234
628.9'6——dc19   CIP

The paper in this book meets the guidelines for
permanence and durability of the Committee on
Production Guidelines for Book Longevity of the
Council on Library Resources.

10 9 8 7 6 5 4 3 2 1

Contents

Acknowledgments

We wish to thank William A. Copper for editing assistance and graphics, Joanne Fox and Dorothy De Mars for typing, and Karen Branson and Steve H. Dreistadt for providing editorial assistance, coordination, and special help in bringing this book to completion. In addition, Tom J. Eager, Paula K. Kleintjes, David L. Rowney, Susan M. Tait, and Robert L. Zuparko assisted in proofreading the text.

The opinions and statements in the chapters of this book are the personal opinions of the authors and do not necessarily reflect the official position of any organization, state or federal agency.

Eradication Concepts

I

Eradication as a Pest Management Tool: Concepts and Contexts

Donald L. Dahlsten, Richard Garcia, and Hilary Lorraine

The tremendous controversy surrounding the 1980–82 Mediterranean fruit fly eradication project in California provided the genesis of this book. During that period a seminar was organized by the graduate students and faculty in the Division of Biological Control and Department of Entomological Sciences at the University of California, Berkeley, to evaluate in depth the concepts and practices involved in eradication programs.

Eradication attempts are usually in response to reports of entry, most often at such points of introduction as ports, borders, and through the mail system. The ease with which exotic pests travel from point to point reflects increased human mobility, and although exclusion has been used successfully in the past, it is becoming increasingly less reliable as a means of preventing pest dispersal. This is the case because controlling the movement of exotic pests by regulating the movement of people is difficult logistically and also introduces the social costs associated with diminished civil liberties and the monetary costs of implementing adequate inspection of passengers, baggage, and cargo. Consequently, much of eradication policy focuses on early detection and the development, selection, and application of eradication measures after the insect has been introduced. For this reason, neither the location of eradication efforts nor their target pests are the central focus of this text, but rather the policy arena in which eradication programs are carried out.

Eradication, as a pest management tool, is based on underlying concepts and existing technologies that are implemented in a policy arena composed of laws and regulations, private and public organizations, government agencies, industry, interest groups, and private consumers. All of these are interdependent and synergistic. That is, existing pest management technologies determine the upper and lower limits for program objectives, and the expectations, demands, and resources exerted by various parties in the agricultural policy arena create the market for developing technology and/or limiting its application. Indeed, the possibilities and the limitations are constantly changing and are the cause and the result of these interactions. Because of this, each participant helps to shape the policy arena. The authors of this volume describe the different ways this shaping process has taken place.

In chapters 2–6, the authors delineate scientific, legal, and institutional approaches to the thorny problem of how to decide if, when, where, and how eradication is a good strategy. They describe a problem that cannot be unraveled simply through available rational decision-making procedures. There is inadequate theory and insufficient data for measuring or even identifying the benefits and costs of any action or inaction. The decision arena is occupied by many competing interest groups attempting to influence a decision permeated by complex and often subjective questions of value, politics, technology, and economics. The outcome, however, is not haphazard and is strongly influenced by a series of biases built into each aspect of the decision-making context.

In chapter 2, John Perkins describes the historical context of modern pest management, tracing the development of pest management strategies as a response to the threefold problem stemming from unbridled pesticide use: resistance, pest resurgence and secondary pest outbreak, and health and environmental concerns. He outlines political, social, and philosophical forces that may be important obstacles to successful eradication efforts and suggests seven questions for gathering the information necessary for a good eradication decision.

Phillip LeVeen, in chapter 3, takes Perkins' discussion a step further, describing the complex process involved in making the kind of evaluation proposed in chapter 2. LeVeen constructs a hypothetical ideal world in which all of the questions posed by Perkins can be answered and the evaluator has access to complete and accurate information on any issue relating to the eradication program. The dilemma discussed by LeVeen is that even if the evaluator could identify and completely describe all of the consequences of the eradication policy, those impacts are difficult to directly compare and weigh. Somehow, values placed on noncomparable impacts must be made comparable in order to decide if the policy is desirable. A standardized unit of measurement, such as a dollar value, is needed. However, LeVeen goes on to sketch several common errors that are made in economic evaluations of pest-control strategies when decision makers assign a dollar value to many noncomparable impacts and then estimate a total value for all the costs and all the benefits of a program. Those errors, reinforced by lack of essential information, bias the evaluation in favor of immediate and tangible results over future and uncertain events and tend to bias the conclusion in favor of the eradication policy.

While the tools of evaluation introduce biases because of the difficulty of assigning comparable values to diverse and noncomparable impacts, the evaluation process itself, as described by Olaf Leifson in chapter 4, takes place in an institutional and asymmetrical context that produces yet another set of biases inadvertently favoring the values and interests of agriculture. Formulation and implementation of policy and evaluation of impact occur within agricultural agencies and rely primarily on information provided by that industry.

The decision is also shaped by the legal context of eradication, described by Charles Getz in chapter 5. This includes promotion and protection of agricultural industry and prevention of the introduction and spread of injurious insect and vertebrate pests, plant pests, and noxious weeds. Within the guidelines of a variety of statutory requirements, administrative experts in the California Department of Food and Agriculture (CDFA) and the United States Department of Agriculture (USDA) formulate detailed policy and create programs for achieving legislatively mandated policy objectives. In this sense decisions about eradication are quasi-legislative. Courts may only examine the legal validity of the decision-making process and must otherwise defer to the discretion of the administrative experts unless there is a clear abuse of discretion or the program is arbitrary and capricious. Policy formulation and program implementation are protected from direct public scrutiny and can be altered only indirectly through legislative change in the agency's broad mandate. Hence, opponents of eradication programs can challenge the validity of the decision-making process but not the goals of specific programs nor the broad policy objectives of the agency.

As Jerry Scribner describes in chapter 6, the challenge process tends to lead to a clash between the short-term management needs arising during a governmental response to crisis and the traditions of our political democracy. These traditions assume that competing political and ideological constituencies will do battle in the market place of ideas. Scribner describes the pivotal and often capricious role of information and misinformation, timing, access to decision makers, structural conflict, personality, and logistics as they influence the decisions of administrative experts in the minute-to-minute course of project management.

Definition of the Problem

Quarantine regulation and government involvement in eradication projects have been justified in terms of economic efficiency and the role of government in providing collective or public goods. The efficiency argument states that society's income will be maximized by excluding unwanted pests and, if exclusion fails, eliminating the pest on a one-time basis is preferable to long-term control strategies. Justification of quarantine and eradication by the government rather than private sector efforts is based on the argument that the free market will underallocate resources to these efforts.

Although justifications based on efficiency and public good are arguably valid in theory, the biases described by the authors badly skew the evaluation and decision-making process toward decisions favoring eradication. The biases work in the following way: Potential pests are rated according to their economic impact. The evaluation of the economic impact is done almost exclusively by the affected industry and its support services. Further, federal and state agricultural agencies rarely develop independent data to determine whether the specific insect poses a significant economic threat, and once a pest is classified as requiring eradication, quarantine regulations mandate exclusion and eradication efforts until the lead agency determines, through an unspecified process, that eradication is infeasible or that efforts have failed. The evaluation of the projected losses that will result from the pest's establishment, a list of control or eradication strategies along with an assessment of their effectiveness, and a statement evaluating the environmental effects of these various options are developed by participating agricultural agencies.

Analysis of the economic impact of eradication projects measures the benefits as the sum of the losses to individual growers, producers, and marketers that would be averted by the eradication project. The analysis of the projected losses rarely includes the effects of elasticity of demand on prices, changes in demand, or shifts in the dynamic equilibrium of the industry. As a result, the benefits of the eradication program may often be overstated. On the other hand, projected program costs are typically understated because they usually include only the immediate outlay for personnel, materials, and equipment and ignore third-party effects, the value of future options, and uncertainty costs. With the exception of techniques using sterilization or male annihilation, information about the available eradication and control strategies is usually generated by chemical producers or via university research grants that are often provided by commodity groups and chemical producers.

Evaluation of economic effects, identification of alternatives, and evaluation of their projected impact take place in an asymmetrical decision-making context. Agencies and industry have full access to the relevant information and can come to a decision through a process that is usually hidden from public scrutiny. In contrast, third-party interests face high costs for information, are often unaware of their possible stake in the decision, and seldom have ready access to the decision-making process.

Once a program is initiated, ongoing evaluation and modification are difficult for the following reasons. Decision makers at all levels must rely on information provided by direct participants, often those with forceful personalities and strong verbal skills. Information that is inconsistent with the opinions and perceptions of an inner circle of managers and advisers is usually filtered. Program managers rely heavily on past action as a guide to future action, and organizational inertia creates pressures for continuing with a course of action once it is chosen. Managers and advisors tend to overestimate the chance of success for their preferred alternative and underestimate the chance of success for competing alternatives.

The pivotal concept underlying eradication policy, unlike control strategies, is reducing the target pest population to zero. While biological control techniques can often maintain a pest population well below the economic threshold level, a residual population remains in order to sustain the parasite or predator population. Chemical control strategies and Integrated Pest Management (IPM) techniques also seek only to maintain the pest population below its economic threshold. The economic threshold for exotic pests, which are considered eradicable, is zero. Because of the zero population criteria, eradication efforts are subject to the law of diminishing returns. In terms of cost per insect, it may be relatively cost-effective to kill 90 percent of a pest population; however, that cost usually escalates with the last 1 percent or so, usually requiring a very costly effort as the eradication effort seeks to achieve 100 percent kill. The same is true of expenditures for detection. Low-level populations can usually be detected and monitored easily; however, detecting these populations with certainty as they diminish becomes increasingly expensive. It is difficult or impossible to compare the exponentially escalating costs associated with the diminishing returns characteristic of eradication efforts against the ongoing costs of controlling a pest below its economic threshold. Indeed, that comparison is not addressed in the evaluation process because the economic threshold for "eradicable" insects is, by definition, zero.

Several patterns emerge from the case studies discussed in chapters 7–17, which reflect the decision biases described in chapters 1–6.

1. Eradication programs are operated by large government agencies, and the administrative components of programs are usually physically far removed from the project site. Consequently, upper management and top-level decision makers often lose sight of the biological and ecological components of the program; supervising at the administrative level becomes the main focus.

2. The importance of understanding the distribution and abundance of the targeted organism in an eradication effort has been de-emphasized. Consequently, project managers have not realized the importance of developing and implementing appropriate procedures for sampling and monitoring the pest's behavior under eradication conditions, and biologically rational criteria for evaluation and action have not been developed.

3. Many important potential invading pests have not been studied thoroughly enough because of inadequate research funding. In the absence of

essential information about the pest's behavior and distribution in the new environment, eradication programs have operated in a crisis-management mode.

4. When eradication projects have been treated as crises, program management has relied heavily on chemical pesticides because of their immediate impact on target pest populations. In many instances, the tendency has been to overkill—higher than necessary dosages have been used, excessive applications have been made, and larger areas than necessary have been treated—leading to an overuse of pesticides.

5. The use of chemicals has enjoyed immense popularity among program administrators as the best approach to eradicating organisms, leading to an overreliance on chemicals and, in many instances, to an attitude of "superiority over nature" (Perkins 1982; see also chapter 2).

6. Very little effort has been made to identify and evaluate the side effects of eradication efforts. Occasionally, chemical monitoring of vegetation, soil, and water has been done to analyze for pesticide residues. Biological monitoring to identify and measure impacts on nontarget and beneficial organisms has been infrequent and when done has been short-term and usually not an integral part of the eradication effort. A recent exception was the 1980–82 Medfly eradication program in California, in which CDFA instituted a study of the side effects of malathion bait spray; however, this was not started until after the spraying had begun.

7. Advice from the scientific community has been sought only when the agencies need support for their programs. Consequently, advisory committees have usually been composed of scientists who agree with the agency's objectives and preferred means. Scientists who recommend alternative means or provide conflicting interpretations of available data and theory are generally not included in the advisory process. A notable exception has been CDFA's recent experimentation with a variety of scientific advisory panels.

8. Information on eradication programs provided to the general public has historically presented only the viewpoint of the administrative agency. That viewpoint is often justified by playing on public fears and relying on narrow evaluations of impact that overestimate the consequences of establishment of the target pest and underestimate the total costs of the eradication effort. Individuals who question the wisdom of eradication strategies or the ability of decision makers to identify and assess all of the important consequences of eradication strategies are often treated with suspicion or ignored. Czerwinski and Isman (1986) suggest a step-by-step procedure for ameliorating some of these problems.

9. Once initiated, eradication programs have lacked appropriate mechanisms for evaluating their activities critically and modifying or even terminating the eradication process when evidence of failure is present.

A central theme emerges: the decision-making process surrounding eradication efforts has led to predetermined conclusions, which, in retrospective analysis, have often led to an inefficient allocation of public money and an inequitable distribution of both the benefits and costs of eradication programs.

Recommendations

The idea of eliminating a pest completely has had great appeal to those involved in pest management. The concept of eradication, however, has generated heated debate between its proponents and those who feel that such an approach is biologically and economically unrealistic. The debate is not new; it has recurred each time an eradication program has been proposed since the beginning of the century when Congress appropriated funds to eradicate the European corn borer, *Ostrinia nubilalis* (Hübner) (Cox 1978). The definition of eradication, however, is uncertain and variable. That is, both the size of the eradication zone and the period of time over which the population of the insect pest must be reduced to zero are variable, depending on the political and economic context of the eradication program (Dahlsten 1986). The certainty with which the pest's absence or presence is measured depends on the efficacy of available detection technologies. The conceptual impossibility of measuring zero population with absolute certainty forces decision makers to balance their need for increased certainty within the context of budgetary constraints. The inherent vagueness of what is meant by eradication, exacerbated by subtle biases that predispose the decision outcome, argues persuasively for expanding the evaluation and decision-making process and for developing criteria that address both efficiency and equity considerations (see DeBach 1964).

The efficiency criterion requires determining if the benefits of a given eradication program will exceed its costs. Although the criterion is simple conceptually, it is difficult to apply rigorously, except in very restricted situations. This is true because the comparison of costs to benefits is determined by the parameters included in the analysis and the time frame used. These are in turn determined by the values assigned to potential impacts and the availability of techniques for identifying and measuring such impacts. The case of eradication poses significant problems because, in addition to operating without essential information, the evaluation process has taken place in a policy environment that has not yet established paradigms that identify important impacts and specify appropriate evaluative methodologies for measuring and evaluating those impacts. Although such an effort is beyond the scope of current knowledge and understanding and, indeed, is beyond the scope of this book, the editors have drawn from the contributions to this volume to identify some important first steps.

Economic evaluation should be expanded to include the following considerations:

Conflicting impacts. The eradication project should be examined for unanticipated and unintended impact on other agricultural interests. DeBach (1964) uses yellow star thistle (*Centaurea solstitialis* L.) as an example. This introduced plant is detrimental to the cattle industry but a benefit to the bee industry. Another example that can be used is the gypsy moth (*Lymantria dispar* [L.]), which can be shown to have a positive influence on net primary productivity in the forest (Mattson and Addy 1975). Other positive effects of the invading species may be the displacement of known pests or possibly desirable changes in the plant community. For example, the gypsy moth could conceivably displace the California oak worm (*Phyrganidia californica* Pack.) in urban areas, and although the gypsy moth has a broader host range than the oak worm, the end result may be swapping one pest for another at no additional cost. Foresters may even find the gypsy moth a positive influence since the black oak (*Quercus kelloggii* Newb.) in California and red alder (*Alnus rugosa* [Du Roi]) and willow (*Salix*) in Oregon and Washington are desired hosts. Although the gypsy moth feeds on conifers at high population levels, it may result in reducing the number of hardwoods in mixed conifer stands and save foresters the expense of using herbicides to control the hardwoods.

Relative potential danger. The case histories included in this volume illustrate that there are great differences in potential for damage and that every pest poses a different hazard. The gypsy moth, for example, has one generation per year and is primarily a shade-tree pest, whereas the fruit flies have multiple generations and attack important agricultural crops. In cases where the relative potential danger is small compared to that of another exotic pest, developing improved methods of control or eradication of the latter one is a better allocation of resources.

Ease of eradication. Those species that escape detection and become well established and widely distributed should probably not be subjects for eradication programs. They pose many hidden costs, which if properly included in a cost-benefit analysis would result in a loss. Hidden costs might include the risk of undetected local populations, a pest's reintroduction, the development of improved detection technologies, increased administration to coordinate a complex program, and public relations to avoid criticism. On the other hand, localized occurrences of introduced pests may warrant eradication, with two caveats: that there are effective traps in place and that control procedures are near 100 percent efficacious at low population levels. The current control practices have been tested and applied against high population outbreaks. Unfortunately, these same techniques are generally used in eradication programs for sparsely populated pests with little knowledge of their impact against low populations. Evaluation of the efficacy of these techniques is extremely difficult in eradication programs because so many factors are involved in a pest's successful colonization of an area. Consequently, failure to detect the presence of a pest for a year or two after an eradication attempt can only be considered circumstantial evidence of eradication. Research must be conducted on methods to insure eradication of the pest at all population levels.

Ease of recolonization. Insects that have already colonized parts of the United States, or any large land mass or continent, probably should not be the targets for eradication programs in other sections of the country because of their potential for recolonization. Insects that can easily recolonize pose a host of hidden costs similar to the ones discussed above. Four recent eradication programs in California have been conducted against insects in this category: the Japanese beetle (*Popillia japonica* Newman) from the eastern United States; the gypsy moth from the East, Michigan, and possibly Oregon as of 1984; the apple maggot (*Rhagoletis pomonella* [Walsh]) from Oregon; and the cotton boll weevil (*Anthonomus grandis* Boheman) from Arizona. Even though individual eradication attempts may succeed, recolonization from bordering states seems inevitable. The high cost of eradication on a recurring basis exceeds the long-term costs of control and cannot be justified on the basis of the efficiency criterion.

Ease of reintroduction. This refers to species that do not colonize an area but may be introduced frequently. The ease with which a pest can be reintroduced presents evaluators with much the same set of considerations as ease of recolonization. In both cases, human behavior plays a large role, and evaluators must take this into consideration. The increased costs of quarantine activities in cases where the pest can be transported by various human activities must be included in the original cost-benefit analysis.

The availability of alternative control strategies. Certain pests, for example, some of the aphid, whitefly, and scale insects, are more amenable to biological control than insects in other groups. The California red scale (*Aonidiella aurantii* [Maskell]), a pest of citrus, was the subject of an unsuccessful eradication program for many years in the Fillmore area of southern California (DeBach 1964). The problem was finally solved by a successful biological control program in the area. The wooly whitefly (*Aleurocanthus woglumi* Ashby) offers another example of a pest for which a chemical eradication program failed and for which a biological control program was a successful alternative (DeBach and Rose 1976, 1977). The alternative to the gypsy moth eradication program in Michigan was the adoption of a policy to spray when defoliation was visible. This was concluded to be economically and environmentally the least costly (Morse and Simmons 1978).

Action guidelines. The inclusion of these considerations in economic evaluation suggests the following guidelines to both researchers and agency administrators. Encouraging the development of population models and basic biological data will aid in estimating the phenomena described above. Information about migratory capacities of the pest and estimates of the probabilities of establishment in the new environment, given effects of harvest, packing, shipping, and marketing, is necessary to determine the characteristics of successful quarantine and for beginning to evaluate the ease of reintroduction, recolonization, and the potential for displacing competing pests.

Whereas the efficiency criterion addresses economic considerations, the equity criterion refers to the distribution of the costs and benefits of eradication and is usually addressed in the political arena. Indeed, an efficient solution can be politically unacceptable because the costs are unfairly distributed and appropriate tradeoffs cannot be developed. In eradication decisions, the affected parties often are not aware of their stake in the decision nor are they fairly represented because of measurement biases in the evaluation process and an asymmetrical decision arena. Hence, information for adequate consideration of distributional issues is not likely to be developed.

The direct monetary costs of eradication programs are borne by taxpayers, whereas the beneficiaries are primarily agricultural and forestry interests. An argument can be made that the benefits are eventually redistributed to consumers through lower prices. But because urban centers are the point of introduction of many exotic pest species and consequently are eradication sites, the health and environmental costs are concentrated in highly populated areas. Hence, the equity criterion should address primarily the distribution of health and environmental costs in these areas. Indeed, most public concern surrounding eradication programs stems from a belief that available information is inadequate and that decision makers are unable to identify and evaluate important long-term consequences to public health and the environment (Lorraine 1984).

The evaluation of impacts on health should be expanded to identify and measure such chronic effects as subtle changes in memory and learning in addition to conventional measurements of such acute effects as carcinogenicity and teratogenicity. Evaluation of environmental impacts should be expanded to regularly include short- and long-term impacts on nontarget organisms.

These recommendations also suggest the expansion of the scope of the cost-benefit analysis to more accurately include measures of social impact, interests of affected groups, and procedures for identifying and notifying all interested parties. Also, because information is costly, incentives must be created for affected interests to generate and supply information that illuminates their gains and losses. This in turn implies mechanisms for introducing symmetry in the decision arena at all levels, for example, through establishing advocates for interested parties and unrepresented values, expanding the scientific and policy advisory committees, or experimenting with policy review committees.

Conclusion

As a result of the California 1980–82 Medfly Program, the relationship between the research community and the agricultural action agencies has been changing rapidly. Program administrators are more receptive to developing a better understanding of basic pest biology, and alternatively, researchers are beginning to address questions that directly respond to the informational needs of decision makers involved in eradication projects, including program managers, environmental groups, and community interest groups. Although this is a good beginning, the budgeting process still results in an allocation of research dollars that is not responsive to developing the kinds of information that are crucial for better decision making about if, when, and how to eradicate an exotic pest. The problem is threefold. Because the research budget of agricultural agencies is subject to the vagaries of the political arena, it is difficult to insure the continuity of effort needed; nor is there the capacity to address long-term issues. We suggest that agricultural agencies, agricultural industries, and community action groups develop legislation that authorizes and requires separate funds earmarked for the following research activities: (1) anticipatory studies of exotic pests with a potential for introduction (as has been done in Australia and Canada), including development of the biological data necessary for estimating the impact on agricultural hosts in the context of current market conditions and available control technologies, (2) studies during eradication programs to learn more about the exotic pest in its new environment in order to provide a better understanding of its behavior as a base for improving detection technologies and developing realistic estimates of damage should it become established, (3) research aimed explicitly at developing less chemically intensive IPM strategies as a backup should eradication fail or as a cost-effective alternative to eradication, and (4) ongoing monitoring of the impact of eradication strategies on nontarget organisms.

Eradication as a pest management strategy is often justified in terms of economic efficiency. The efficiency argument states that societal income will be maximized by excluding unwanted pests, and if exclusion fails, eliminating the pest on a one-time basis is less costly than long-term control strategies. The problem identified in chapters 2–6 and confirmed by the case studies in chapters 7–17 is that biases exist that predispose the decision-making process toward eradication. In eradication decisions, values play a large role, and science can only partially resolve the uncertainties relating to the impact of a given action. The benefits and risks can range from insignificant to immense, and there will always be disagreement over how to weigh the evidence and place a value on the outcome. Arguable, too, is the question of whether the values of the general public are being represented and whether the consequences important to the public have been properly identified. In this context, developing ways to compensate for these biases is a first step toward improving the decision-making process with respect to both efficiency and equity. As the contributors to this volume suggest, the changes necessary for making appropriate decisions require both wide dissemination of information and decision making that incorporates a variety of interests and values at all stages.

Just as values strongly influence eradication decisions in the policy arena, they have also shaped the discussion presented here, and the introductory chapter to this exploration of eradication would not be complete without an explicit comment about the values of the editors. The editors believe that eradication may have clear short-term benefits in some cases—for example, as a stopgap effort until alternative control techniques are available or in cases where pests can be detected at very low population levels and efficacious, environmentally sound techniques are available. Because of increasing contact and exchange throughout the world, introductions of exotic pests will take place with increasing frequency, and the editors believe that eradication will not be a rational or a wise long-term solution as a pest management strategy. In the long run, developing alternatives to eradication will be economically and environmentally less costly. In the absence of complete and unequivocal information to make such calculations, however, the recommendations discussed here are offered as contributions to an ongoing dialogue designed to reach better decisions about if, when, and where to use eradication as a short-term solution.

Literature Cited

Cox, H. C. 1978. Eradication of plant pests—pros and cons. *Bull. Entomol. Soc. Amer.* 24:35.

Czerwinski, C., and M. B. Isman. 1986. Urban pest management: Decision making and social conflict in the control of gypsy moth in west coast cities. *Bull. Entomol. Soc. Amer.* 32:36–41.

Dahlsten, D. L. 1986. Control of invaders. In *Ecology of biological invasions of North America and Hawaii*, ed. H. A. Mooney and J. A. Drake, 275–302. New York: Springer-Verlag.

DeBach, P. 1964. Some ecological aspects of insect eradication. *Bull. Entomol. Soc. Amer.* 10:221–24.

DeBach, P., and M. Rose. 1976. Biological control of woolly whitefly. *Calif. Agric.* 30:4–7.

————. 1977. Environmental upsets caused by chemical eradication. *Calif. Agric.* 31:8–10.

Lorraine, H. 1984. The 1980 medfly crisis: an analysis of uncertainty management under conditions of non-routine problem solving. Ph.D. diss. Univ. of Calif., Berkeley.

Mattson, J. J., and N. D. Addy. 1975. Phytophagous insects as regulators of forest primary production. *Science* 190:515–22.

Morse, J. G., and G. A. Simmons. 1978. Alternatives to the gypsy moth eradication program in Michigan. *Great Lakes Entomol.* 11:243–48.

Perkins, J. H. 1982. *Insects, experts, and the insecticide crisis.* New York: Plenum.

2

Eradication: Scientific and Social Questions

John H. Perkins

Eradication or annihilation of insect pests has long been a vision that appealed to victims of the pest and to pest-control scientists. The essence of eradication's popularity lies in an alluring simplicity: The offending insect is completely removed; then not only is all current damage eliminated but no future harm can come from offspring of the current generation of the pests. In brief, eradication holds out the promise of *total* and *perpetual* relief from torment.

Examples of many unsuccessful and a few successful eradication plans dot the history of early twentieth century entomology in the United States:

—The Commonwealth of Massachusetts attempted to eradicate the gypsy moth (*Lymantria dispar* [L.]) from its domain in the late 1800s (Forbush and Fernald 1896);

—USDA and several states cooperated in an effort to suppress the movement of the European corn borer (*Ostrinia nubilalis* [Hübn.]) into the vast acreages of the Corn Belt in the 1920s (Dunlap 1980);

—A small infestation of Mediterranean fruit fly (*Ceratitis capitata* [Wied.]) was eradicated from Florida in 1929 (Metcalf, Flint, and Metcalf 1951, 760);

—Governor Huey Long of Louisiana was persuaded to endorse a complete moratorium on cotton production in 1932 in order to eradicate the boll weevil (*Anthonomus grandis* Boheman) (Williams 1970, 530–33). The moratorium was not conducted;

—Professor Clay Lyle, President of the American Association of Economic Entomologists, urged massive campaigns to eradicate a variety of

An earlier version of this paper was delivered to the symposium on Bioethics in Pest Management at the joint meeting of the Entomological Society of America and the Entomological Society of Canada, Toronto, Ontario, December 2, 1982. I wish to thank Drs. Richard T. Rousch and William Uphold for valuable comments on this manuscript. Part of the support for the research came from the National Science Foundation (SOC 76-11288), for which I am deeply grateful. Ms. Sharon Tucker, Ms. Joylan Netter, and Ms. Paula Butchko were very helpful in typing the manuscript and in other ways.

agricultural and public health pests with the tremendous killing power of the synthetic organic insecticides invented in the 1940s (Lyle 1947).

For a variety of biological and social reasons, proposals and campaigns to eradicate insect species over large areas of land were disappointing. The only real successes were in cases in which the target insect was confined to small areas or to a sharply limited breeding habitat. Insect control technology was limited, and the economic and political obstacles to massive campaigns proved overwhelming to the eradication proponents.

Failures and difficulties in early efforts were not sufficient to diminish the appeal of eradication as a concept for insect-control technology. Most of modern science is predicated on a notion of *control* of natural processes. For entomology, eradication is the ultimate in control. The concept of insect control through eradication is so alluring to ambitious and imaginative scientists that it was and remains irresistable. Failure to achieve eradication in a venture was merely a signal to head back to the laboratories for further study, not an indication that eradication might be too difficult technically, impractical on economic and political grounds, or undesirable on philosophical grounds.

Eradication as a concept of insect control emerged with renewed vigor in the late 1960s and through the 1970s. Entomologists pursuing eradication in its contemporary guise developed a systematic theory of how mutually reinforcing suppression techniques could be applied over thousands of square kilometers to achieve total annihilation of a particular insect species. For selected insects, these entomologists developed comprehensive biological and bureaucratic protocols that, when implemented, could theoretically reduce the target population to zero. Elsewhere I have called this new strategy Total Population Management or TPM (Perkins 1982). As a strategy for insect-control research and operations, TPM was not always dedicated to eradication as a goal, but its origins were deeply rooted in efforts to eradicate the screwworm fly (*Cochliomyia hominovorax* [Coquerel]) and the boll weevil.

Modern eradication theory, in the form of TPM, emerged in response to a crisis involving the use of such synthetic organic insecticides as DDT (dichloro-diphenyl-trichloroethane). The new insecticides were highly successful, but their continued use frequently led to resistance of the target population, destruction of natural parasitic and predatory insects with accompanying outbreaks of insects previously controlled by the destroyed natural enemies, and unacceptable damage or hazards to the environment and human health.

TPM was not the only response to the crisis surrounding the use of synthetic insecticides. Integrated Pest Management (IPM) was an alternative strategy developed by entomologists at the University of California and elsewhere. IPM explicitly disavowed eradication as a concept of control except for extremely limited situations in which the pest species was confined to a small geographic area. Instead, IPM was a strategy for research and control that aimed for management of a population of pests that was presumed to be permanently present. Proponents of TPM and IPM were frequently at loggerheads over policy questions concerning entomological research priorities. Rivalry between the two strategies often made understanding either of them a difficult task.

Most of the case studies recounted in this volume date from before TPM was articulated sufficiently to have influenced them. Nevertheless, future eradication programs in the United States and elsewhere are sure to be influenced by the emergence of both TPM and IPM. Probabilities of success for future eradication efforts are variable but likely to be low. Reasons for this pessimistic outlook are based on a combination of technical and social problems, each of which can be outlined in terms of the tenets and assumptions of TPM as a strategy for research and operations. This chapter outlines the nature of those problems in the context of the historical development of the insecticide crisis, TPM, and IPM.

Insecticides and the Insecticide Crisis

Economic entomology was a profession and academic discipline by the last quarter of the nineteenth century. Early practitioners recognized a number of general tactics for controlling pest insects: biological control, chemical control, sanitation and habitat management, physical destruction of individual insects, host-plant resistance, and quarantines to prevent a destructive species from entering an area. Despite the successes of the early entomologists, the power of their remedies frequently left much to be desired. With a few exceptions, no one tactic for insect control was sufficiently useful to raise it to dominance among all of the known methods of control.

Paul Herman Mueller, a Swiss chemist in the employ of the J. R. Geigy, Co., unknowingly forced a profound change on all of economic entomology in 1939. In September of that year, Mueller discovered the insecticidal properties of the compound DDT. He had begun his search with the rather narrow objective of finding a compound that could effectively mothproof woolen clothes and that was inexpensive, durable, and with low toxicity to humans and other warm-blooded animals. DDT met the criteria established by Mueller, but it also established a basis by which chemical control of pest insects became the foundation for almost all pest suppression activities (Perkins 1978b).

The early history of DDT was completely intertwined with the events of World War II. In the United States during the war, limited amounts of DDT were made available for testing in agriculture, but most experimentation and virtually all production of the compound was directed toward military uses. Preoccupation with the war notwithstanding, Dr. Sievert A. Rohwer of USDA reported in 1945: "We feel that never in the history of entomology has a chemical been discovered that offers such promise to mankind for relief from his insect problems as DDT. . . . This promise covers three chief fields: public health, household comfort and agriculture" (Rohwer 1945).

The early successes of DDT taught an important lesson to the en-trepreneurs of the chemical industry: synthesize and screen for toxicity as many chemicals as you can; other success stories lie hidden amid the myriad molecules that the chemist can imagine and synthesize. Millions of dollars were poured into the search for more magic chemicals, and by the 1960s the number of successes was clear justification for the effort. Aldrin, BHC (benzene hexachloride), dieldrin, chlordane, heptachlor, endrin, carbaryl, diazinon, parathion, TEPP (tetraethyl pyrophosphate), and many others joined DDT in the ranks of chemical tools for killing unwanted insects. The chemical industry had enjoyed limited success in markets for insect control before DDT, but afterward its fortunes rose markedly. Manufacturers of insecticides saw their sales jump from $9.2 million in 1939 to $174.6 million in 1954 (Bureau of the Census 1957). Production of DDT, for example, rose from about 10 million pounds in 1944 to a peak of 188 million pounds in 1962–63 (Economic Research Service 1969).

The chemical industry was not the only economic interest group to benefit from the technological prowess of the new chemicals. Farmers, ranchers, and foresters cut their losses and raised their profits by adopting the new materials. Apple producers, for example, faced annual losses of 20 percent to 95 percent from codling moth (*Laspeyresia pomonella* [L.]) if it went uncontrolled (Metcalf and Flint 1939, 594, 599). Lead arsenate, the compound of choice for codling moth from the early 1900s to 1950, could reduce losses to about 15 percent. Use of DDT allowed the orchardist to cut the wormy fruit to as low as 1–2 percent or less (Baker 1952). Predictably, DDT replaced lead arsenate in apple production. Similarly, cotton growers plagued with the boll weevil began using BHC, toxaphene, endrin, and aldrin and enjoyed great success compared to the times when they had relied on arsenate and preventive farming practices. Destruction of boll weevils with the chlorinated hydrocarbon insecticides was so powerful, in fact, that cotton producers after the late 1940s began to drop their earlier practices of wide-row planting, low fertilization, and little irrigation—all of which diminished the ability of the boll weevil to reproduce. Narrower rows, higher fertilization, and more irrigation all increased yields, but only with an umbrella of insecticidal protection against the boll weevil, which rapidly reproduced under the altered cultural practices (Newsom 1974).

As farmers became ever more dependent upon insecticides for insect control in the late 1940s and early 1950s, other control methods that had been known and used for years dropped by the wayside. Biological control became less effective as the toxic chemicals destroyed predatory and parasitic insects, as well as pest insects. Cultural control methods, as exemplified by the cotton producers, became technologically obsolete. Chemicals that had enjoyed high use rates before the invention of DDT, such as the arsenicals, also were retired as obsolete. Chemical control based on the miracles of the synthetic organic insecticides became the paramount insect-control tactic. With the demise of other tactics, chemical control became the premier strategy for insect control in North America.

If the initial successes with the new, synthetic organic insecticides had simply repeated themselves year after year, the chemical control strategy would still be the unquestioned king of the insect-control technologies. But it is undisputed within the profession of entomology that the euphoric triumph over insects heralded by the new chemicals gave way first to a growing despair and later to a creative set of responses in the form of new strategies for insect control. Motivation for the invention of new paradigms for research and practice came from the now-familiar three scourges of overreliance on chemicals.

First, the new insecticides were highly efficient in selecting for resistant populations of pest insects. Resistance was recognized as a phenomenon of natural selection of genetic variants before DDT was widely used, but the extent and scope of use of the synthetic organics created more resistant populations than had ever been recognized in earlier times. Thirteen cases of resistance had been documented by 1945, but 124 new cases were identified within the following fifteen years (Brown 1961).

Second, destruction of natural enemies of pest insects led to the counterintuitive result of pesticide treatment of a field causing a pest population *larger* than what existed before treatment or *larger* than what would have developed in the absence of treatment (Smith and van den Bosch 1967). "Resurgence" was the term applied to cases in which there is a quick reoccurrence of the pest that was the original target of the insecticide, whereas "secondary pest outbreak" described the situation in which an insecticide applied for a particular species allowed a different, previously innocuous species to grow to a population size large enough to cause damage. The latter phenomenon was an especially startling result to entomologists unfamiliar with the actions of natural enemies because it amounted to insecticide-induction of a pest problem. The means of relief became the source of the problem.

Third, health and environmental problems were associated with the use of the new insecticides. Ironically, the advent of the synthetic organics was at first welcomed because it was thought that the new compounds would, as a class, be less harmful to people than the arsenicals that were widely used before 1939. Replacement of the arsenicals by the newer compounds mooted the arguments that had raged in the 1920s and the 1930s over the dangers of arsenicals (Whorton 1974). Unfortunately, the synthetics were soon under rapid fire from members of the health professions. The Special Committee to Investigate the Use of Chemicals in Foods and Cosmetics served as an investigative arm of the congress from 1950–52 (U.S. Congress 1952). The Committee's conclusions led to a strengthening of the Federal Food, Drug, and Cosmetic Act in the regulation of pesticide residues on food and feed (68 Stat. 511–17). Rachel Carson's *Silent Spring* (1962) made the more general argument that the new synthetic pesticides in some cases constituted a grave danger to humans and wildlife, even if used as directed.

This chapter is not the time or place to examine the details of each of the major problems associated with the new insecticides. Suffice it to say that each problem provided a disincentive to rely on the chemicals for insect control and an incentive to create alternatives. A careful examination of the timing of each of the problems indicates without a doubt that the first two, resistance and destruction of natural enemies, were sufficient to launch professional entomologists on a voyage of exploration in search of control strategies that would not result in resistance and destruction of natural enemies. The general public is far more familiar with the environmental disputes over insecticides, but those arguments had a secondary role in the invention of new insect-control strategies. Although most entomologists believed that Rachel Carson had presented at best a biased story, she was probably of considerable assistance to them by creating a climate that was conducive to the appropriation of more research dollars. For example, the USDA research budget in entomology rose from $11.2 million to $16.9 million between 1964 and 1965 (U.S. Congress 1965, 737–38; 1966, 533–34). *Silent Spring* created noisy but lucrative entomological politics.

Two major alternative strategies for insect control emerged during the 1950s and 1960s: TPM and IPM. Each was designed to meet the crisis caused by resistance, destruction of natural enemies, and environmental health hazards. Each developed from efforts to control specific insects, and each was elaborated into a general theory that transcended its specific origins. Each strategy gained followers within the entomological community, and followers of each designed their research programs around the general principles of the strategy. Significantly, each strategy had problems getting launched. In many ways, TPM and IPM were rivals, but at the same time each was attempting to win entomologists away from research designed solely around the chemical control strategy.

Strategic Inventions in Entomology: TPM and IPM

TPM

TPM is my designation of the set of ideas and practices originated by Edward F. Knipling of the USDA and his colleagues (Perkins 1982, 97–126). A

comprehensive research strategy designed to create new insect-control technology to relieve entomologists and farmers of the insecticide crisis, it was
also the foundation of a modern and sophisticated eradication concept.

Knipling's strategic invention, TPM, represented the convergence of two
different research pathways: insecticides and the sterile-male technique.
Knipling played highly important roles in both areas.

The Geigy Company first introduced DDT into the United States. Representatives of the Swiss firm delivered samples of the new compound in 1942
to the Bureau of Entomology and Plant Quarantine at USDA. The Department of Agriculture had already launched a massive program to find substitutes for rotenone and pyrethrum, both of which had become scarce
because of the war. In addition, USDA had assigned its Orlando, Florida,
laboratory to work on problems of military importance. Knipling was a
research leader at the Orlando facility when the first samples of DDT arrived
from Washington, D.C. Within six months, the Orlando laboratory had
demonstrated conclusively that DDT was an extraordinarily useful compound against insects affecting man and made recommendations for its use
to the armed forces (Perkins 1978b).

Successful developments around DDT and other synthetic organic insecticides helped create a new avenue of research for Knipling and his colleagues. Through 1943, for example, fewer than one-third of Knipling's
publications dealt with investigations on insecticides. But from 1947
through 1953 over one-half of his published papers were on insecticides.
Knipling followed the development of resistance to the new insecticides
closely and was clearly concerned about the toxicity problems posed by the
chemicals. Nevertheless, his experiences with the new chemicals were
largely positive, and he defended their proper uses in public addresses
(Knipling 1953). He understood the weaknesses associated with the use of
insecticides at an early date, but he was a staunch defender of using them
when and where they worked. His other lines of scientific investigation
gave him a way to combine the uses of insecticides with other forms of
suppression techniques so as to obtain the strengths of insecticides with
fewer of their liabilities.

Knipling's second major line of successful scientific research was the
sterile-male technique for insect control. In 1937, he and several of his
colleagues were engaged in efforts to control the screwworm fly in Texas
and in the southeastern states. Investigations showed it was possible to raise
larger numbers of the insect artificially than occurred in nature, where the
species was quite rare. Knipling reasoned that if large numbers of artificially
reared males could be sterilized, then their release into the field could lead to
a drastic reduction in the number of successful reproductive events in wild
females. Sufficient numbers of released sterile males could theoretically lead
to eradication of the species from a particular area (Perkins 1978a).

World War II interfered with the plans to develop the sterile-male technique, but Knipling returned to the concept after the war was over. By 1952, laboratory and field experiments conducted by Raymond C. Bushland demonstrated that the technique had promise. A larger-scale experiment on the isolated island of Curaçao demonstrated a complete triumph of the sterile-male technique—the island was completely rid of the obnoxious, dangerous screwworm fly in 1954–55 (Baumhover et al. 1955). Use of the technique led to the eradication of the fly from the southeastern United States in 1958–59 and the substantial suppression of it over the vast grazing lands of Texas starting in 1962 (Scruggs 1975).

Successes obtained with the sterile-male technique plus the considerable positive experience with insecticides led Knipling to construct a formal theory of the principles for insect control by the mid-1960s. He presented his theory in the Founder's Memorial Lecture to the Entomological Society of America in 1965 (Knipling 1966). Knipling argued that the core of a sound system of insect control was based upon attack against the total population of the pest, not isolated pockets occurring in individual fields. He also argued that single suppression techniques were subject to limitations but that combinations of different techniques could compensate for the individual weaknesses and allow the combined package to overcome the law of diminishing returns. Of particular importance, he argued, were the possibilities of combining the use of insecticides with the sterile-male technique. Insecticides were efficient in killing insects when the populations were large but inefficient when the numbers were small. The sterile-male technique, in contrast, was inefficient in killing large populations but highly effective when numbers of individuals were small. A sequential use of insecticides and the sterile-male technique could reduce a large population to a small one and then continue the reduction to a miniscule population or zero. In the extreme case, eradication was possible in theory and, as demonstrated by the use of the sterile-male technique against the screwworm fly, also in practice.

Knipling's personal experience in research had been with insecticides and the sterile-male technique, and his first discussions of the TPM strategy emphasized the use of those particular technologies. He expanded his strategy over time, however, to include various combinations of the male annihilation method, use of poisoned baits, use of pheromones and kairomones, incorporation of new types of chemicals such as juvenile hormones, and uniform practices by farmers of various cultural methods of insect suppression (Knipling 1979). He was also keenly aware of the utility of such legal methods as quarantine and the destruction of noncommercial planting to eliminate insect breeding grounds. In other words, TPM was a comprehensive strategy that allowed any entomologist to incorporate a particular suppression technique systematically into a package that was not subject to the limitations of every suppression method used individually. Moreover, Knipling argued, TPM afforded the possibility of eradication as a permanent solution for particular species in particular areas. He never believed that eradication was useful as a goal in every case, but he advocated its use if a chance for success existed.

Movement of TPM from just a theory to a strategy guiding research was difficult because each experiment designed under the strategy was expensive. The total population of an insect species cannot be attacked over thousands of square kilometers without a well-trained crew of experimentation monitors, extensive use of supplies, and the cooperation of every landholder in the eradication zone. Great efforts must be invested in education about the problems to be attacked, the methods to be used, and the benefits to be gathered if the effort is successful. Costs of implementing a TPM experiment come in the form of monetary expenditures for salaries and material goods plus sociopolitical efforts needed to achieve compliance by all residents in the area. In this very practical manner, therefore, TPM was markedly distinguished from IPM. In the latter strategy, an individual experiment could be launched with far fewer resources.

Only a handful of exercises inspired by the TPM strategy have been launched. Perhaps the most famous have been the two efforts against the boll weevil, the first in Mississippi (1971–73) and the second in North Carolina and Virginia (1978–80). Other efforts have included the oriental fruit fly (*Dacus dorsalis* Hendel) in the island of Rota, the codling moth (*Laspeyresia pomonella* [L.]) in the Pacific Northwest, and the pink bollworm (*Pectinophora gossypiella* [Saund.]) in the San Joaquin Valley of California. These experiments and the sometimes controversial interpretation of their results have all been described elsewhere in some detail. Suffice it to say here that the efforts to eradicate the boll weevil engendered much controversy (Perkins 1980, 1982), the attack on codling moth was technically successful but economically impractical (Proverbs, Newton, and Logan 1977), the effort against the oriental fruit fly was remarkably successful (Knipling 1979, 433), and the success of the attack on the pink bollworm is difficult to judge (National Academy of Sciences 1975; Graham 1978). Another effort inspired by TPM is the ongoing effort to suppress the screwworm fly in Texas and Mexico with the sterile-male technique alone. This program must be judged a successful venture although it has had a few lapses that were corrected (Knipling 1979, 348–50).

IPM

In North America IPM had its origins primarily in the work of Ray F. Smith and his colleagues at the University of California. Many other entomologists—A. D. Pickett and his colleagues in Nova Scotia, for example—

also made important advances. Nevertheless, it was Smith and his colleagues who articulated a coherent theory of integrated control in the late 1950s and demonstrated that the principles could be used to design research leading to practical solutions of insect problems in the field.

IPM represented the melding of two previously separate research lines: biological control and the use of insecticides based upon ecological understanding of the pest species (Perkins 1982, 67–81). Classical biological control involved the search for parasitic and predatory insects in foreign lands, followed by importation, release, and establishment of the species to control a pest species that had been introduced at an earlier time. Successes with classical biological control were based upon the principle that many species of herbivorous insects generally occur at low enough population levels to not be considered a pest. The regulating agents are predatory and parasitic insects. When the herbivorous species is introduced into a new area without its accompanying natural enemies, its population explodes and the organism becomes recognized as a pest. Reassociating the pest with its natural enemies, if successful, restores the population regulatory factors, and the pest population drops to, and remains at, a low level.

Smith was well aware of the successes of biological control because of his location at the University of California, a major site for biological control research. In addition, his personal research involved developing uses of insecticides based upon the ecology of the pest. In the late 1930s, Smith's mentor, A. E. Michelbacher, noted that larvae of the alfalfa butterfly (*Colias eurytheme* Boisduval) were often parasitized by the native parasite, *Apanteles medicaginis* Muesebeck. Further work by Michelbacher and Smith indicated that a farmer did not necessarily have to treat his alfalfa with insecticide if the parasitic wasp was sufficiently active (Michelbacher and Smith 1943).

In 1946 Smith used this knowledge about the ecology of alfalfa butterfly to organize farmers in the San Joaquin Valley into pest-control districts. They each paid a fee based on their number of acres and hired an entomologist to scout their fields and render an expert opinion about whether the fields needed treatment (Hagen and Smith 1947). In the 1950s Michelbacher dubbed this use of insecticides based upon solid ecological research as "integrated" control, meaning that the use of chemicals was integrated with the reliance on natural biological control (Michelbacher and Bacon 1952).

The spotted alfalfa aphid (*Therioaphis maculata* Buckton) invaded California in 1954, and farmers reacted by using heavy treatments of insecticides. Resurgence plus resistance set in within two years. Smith and three other colleagues initiated research using the principles of integrated control. They were remarkably successful in developing a treatment relying on low doses of demeton plus the use of imported natural enemies of the spotted alfalfa aphid. Their success with this one insect would have been significant, but they expanded their triumph by articulating a general theory of integrated control. Their paper, "The Integrated Control Concept" (Stern, Smith, and Hagen 1959), is now generally cited as the paper in which the IPM strategy was articulated.

Conceptual elaboration of IPM came during the 1960s. Despite a variety of phrasings, the foundation of IPM was that the core of sound technology for insect supression methods, including insecticides, could be used, but the ancillary techniques should be used in a way that did not destroy or negate the core supression methods. Research in the IPM tradition was accordingly oriented toward developing an understanding of how the pest insect reproduced in the face of natural enemies and different plant varieties; heavy emphasis was also given to investigations on the biology of natural enemies and plant breeding.

An implicit assumption was made in most IPM work: the target of the technology was a population of pest insects in a particular field. Researchers believed that individual farmers would need expert advice or scouting from entomologists in order to decide on a field-by-field basis whether the pest population had passed an economic threshold and thus merited treatment. An important corollary of the IPM strategy, therefore, was that a small population was not a "pest." Researchers inspired by these principles generally excluded eradication of the pest species except when the pest was newly introduced into an area and still confined to a small geographic distribution.

Translating this research strategy into actual field studies was a complex process. IPM theorists wanted a nationwide, coordinated effort on many crops that could produce a series of new insect-control technologies that in turn could significantly affect the dependence of agricultural producers on insecticides. Carl B. Huffaker, a colleague of Ray Smith's at the University of California, and Smith emerged as the leaders of a large project designed to develop IPM technologies for cotton, soybeans, stone and pome fruits, alfalfa, citrus, and pine bark beetles. The Huffaker Project, as it came to be known, involved over three hundred researchers from eighteen universities and parts of USDA. These researchers worked from 1972–78, spent about $13 million, and produced over 250 scientific papers. Details of the origins and impact of the Huffaker Project are given elsewhere (Perkins 1982, 61–95, 142–53), but it is important to note that it was the largest single coordinated entomological research program ever designed and conducted outside of the USDA structure.

Translation of research results into practical, available technology that can be used by farmers remains a difficult problem for IPM entomologists. Some significant advances have occurred, such as practices for apples in North America, cotton in parts of Texas, and soybeans in the Southeast (Office of Technology Assessment 1979, 35, 39, 56; Whalon and Croft 1984).

Advances in computer software development are having a positive impact on the ability of some farmers to use IPM (Welch 1984). IPM as an idea and the Huffaker Project as actual experimentation have had important effects on the thinking of entomologists. The scope of this chapter does not allow a detailed discussion of the use of biologically based computer models in managing agricultural systems. An overview of the two basic categories can, however, illustrate the potential that such technology could have for the future.

Strategic models are used in addressing long-term management questions, whereas tactical models are used for the day-to-day management of crops. Wilson et al. (1987) provide a brief account of the features of strategic and tactical models and the philosophy behind the development of each.

Strategic models are usually structured to provide a greater understanding of how various components within a crop system interact. They are often extremely useful in delineating what we do and do not know about a particular system, thus providing a strong focus for formulating a plan of research. Strategic models are also helpful in addressing such long-term questions of management as whether an early season or a late season control program would result in greater farmer profits over several years. Strategic models vary considerably, the complexity and structure determined largely by the degree of biological understanding inherent in the models involved, the level of resolution at which the system is being addressed, and the particular modeling approach favored by those writing the computer code. Strategic models are rarely used by crop managers except when in concert with scientists for demonstration purposes. Their structure is not sufficiently flexible to enable their use in addressing within-season questions of management.

In contrast, tactical management models, although few in number, are increasingly being used in the day-to-day management of crop systems. Such models have in the past relied heavily on information generated by strategic models. The most widely used tactical model is the SIRATAC program developed for managing cotton in Australia (Hearn et al. 1981; Ives et al. 1984). SIRATAC is currently used to manage approximately 25 percent of Australia's total cotton acreage and has been shown to reduce dependency on insecticides by as much as 40–60 percent without a concomitant drop in crop yield or quality. As distinct from the strategically oriented models, SIRATAC and other tactical management models make less use of complex biological algorithms and greater use of simpler statistically generated functions. Although such equations may contain relatively little biological clarity, they nevertheless enable the growth and development of the cotton crop and its dominant pests to be mimicked at a level of resolution sufficient for management of the crop and the associated pest species.

The development of tactical management models is considerably less

advanced than that of strategic models. One reason for this is the requirement that tactical management models mimic conditions faithfully on a
field-by-field basis, whereas strategic models often only mimic the general
trend and magnitude of events. A considerably greater effort is thus required in developing a tactical management model. One might consider this
difference to be analogous to the distinction between model verification
and model validation. This distinction aside, however, a good strategic
model can often serve as the basis for developing a sound tactical management model with a considerable savings in time compared to building a
tactical management model from scratch.

Scientific Questions in Eradication Efforts

Advocates of eradication through a TPM strategy have engaged in vigorous
debate with proponents of IPM and chemical control over the proper research strategy for economic entomology. Both scientific and social issues
are at stake, but little explicit attention has been given to the latter. For
analytical purposes, I will outline separately the major scientific questions
about eradication efforts. It is important to realize, however, that the answers to these concerns must always be given based on a particular species of
pest on a single or limited number of hosts in a specific region of the world
for a particular clientele of pest controllers. It is not possible to disentangle
the scientific issues from the social questions within the context of a particular eradication proposal. We can separate the two types of issues only for
abstract, analytical purposes.

The questions listed below identify scientific issues of utmost importance
to eradication efforts. Each question can be addressed on the basis of data
from orthodox entomological research methods. But reaching interpretations and conclusions from the data is more complex because it often
requires the investigator to make assumptions or draw inferences from
incomplete data (see, for example, Perkins 1982, 130–39). Social issues frequently affect the type of assumptions and inferences made by a scientist.
Making scientific judgments, in short, is not entirely an objective, al-
gorithmic process in which data from experiments lead one inexorably and
unambiguously to one, and only one, scientific conclusion. The potential
for multiple interpretations of the same data makes science interesting, but
ambiguity of interpretation also contains the potential for serious policy
pitfalls. The process of public argumentation about assumptions and interpretations enables the scientific community ultimately to make decisions
that enjoy the support and confidence of the scientific community as a
whole.

1. *How significant is the "insecticide crisis" in particular cases?* Entomologists

acknowledge the general validity of the three components of the insecticide crisis: resistance, destruction of natural enemies, and unacceptable hazards to the environment and human health from the use of insecticides. Justifications for research in both TPM and IPM were frequently couched in terms of efforts needed to avoid the collapse of insecticide-based control technologies. For example, proponents of boll weevil eradication programs and of integrated pest management methods of boll weevil control both alluded to the insecticide crisis as a major justification for their new research programs.

There is no reason to doubt the truth of the insecticide crisis as a phenomenon occurring with many insects attacking many hosts in various regions of the world. As a principle of applied entomology, the concept of the insecticide crisis has validity that all students of the subject should understand. Nevertheless, research policy for particular insect-control problems must be based on a finer and more specific understanding of the insecticide crisis as it affects a particular insect in a particular region.

Proponents of eradication must be especially precise in using the reality of the insecticide crisis to justify massive, area-wide, expensive programs of research and operations. It is not sufficient to defend multimillion dollar exercises on the grounds that (a) some insects have become resistant to some insecticides, (b) some uses of insecticides have ignited uncontrollable secondary pest outbreaks and flarebacks by destroying natural enemies, and (c) some insecticides have created unacceptable environmental and human health hazards. Instead, the justification for any eradication proposal must, at a minimum, explain how each of the three crisis factors plays a role in the chemical efforts to control the target pest in the particular region scheduled for the eradication effort. Failure to provide this sort of scientific analysis indicates that not enough is known about the biology and natural history of the intended eradication target to support the eradication effort.

2. *What scope of attack is needed for reliable and acceptable insect-control technology?* The chemical control strategy was based on individual farmers treating individual fields or animals, and IPM researchers have generally continued that scope of attack against pest insects. Eradication efforts as envisioned in TPM efforts stand in stark contrast to the chemical control and IPM strategies because the former are based on an attack against the total population over a wide area. Enormous consequences follow from TPM exercises using a wide scope of attack. Of foremost importance are (a) needs for improved knowledge about the requirements of the target species for breeding, feeding, and resting habitats outside the boundaries of managed fields and (b) precise coordination among all landowners in the target area so that the control protocol can be uniformly applied. Ensuring uniformity among all landowners implies land-use control, a politically controversial concept in the United States and other market economies.

Little attention has been given to developing principles to guide choices on the scope of attack against a pest (Roush, personal communication, 1988). At a minimum, such guidelines should provide criteria for determining whether sufficient knowledge is at hand to justify the heavy expenses and social disruption that accompany eradication endeavors. Gaps in the understanding of the target pest's reproductive biology, feeding and migration habits, survival mechanisms during adverse times, and relationships with hosts and natural enemies may lead to an unwise decision to launch an eradication effort. Alternatively, missing knowledge may also lead to a failure to initiate either a wide-area attack or eradication effort that might pay enormous economic and environmental benefits. In either case, the lack of knowledge creates a situation in which inappropriate or suboptimal technological decisions are made.

3. *What accommodations need be made for the limitations in the detection methods of ultralow populations?* A reliable capacity to detect extremely low populations of a pest species is necessary for an eradication endeavor. It follows that highly sensitive detection methods are far more important for eradication and TPM than for IPM research. For extremely low populations, use of pheromone wicks or baits in traps has been the foundation of the most sensitive detection methods. Data from pheromone traps indicating ultralow population densities were crucial to the two most important eradication efforts inspired by the TPM strategy: the Pilot Boll Weevil Eradication Experiment (1971–73) and the Trial Boll Weevil Eradication Program (1978–80).

Eradication efforts, even when aided by the latest in pheromone traps, are always at a disadvantage because the absence of anything cannot be directly proven. It is possible only to state that given a particular level of detection sensitivity the entity is or is not present. No detection may indicate that eradication has occurred, but it may also mean that the population is below the sensitivity threshold of the detection method. A period of waiting to allow reproduction and growth of the population is the only way to distinguish between the two possible data interpretations. If eradication was achieved, then waiting will continue to result in no detection. If the population was simply below the sensitivity threshold, waiting will eventually result in detection as the population recovers to the detectable level.

Calculation of the sensitivity threshold is difficult for conceptual and technical reasons. An accurate empirical estimation requires perfect knowledge of the census of an insect population in an area. All census information requires a particular mode of detection, yet all detection modes are imperfect. It is therefore conceptually impossible to have an absolute empirical estimate of detection efficiency for a natural population. The best that can be achieved are estimations of trapping efficiency based on such statistical methods as release and recapture of marked individuals. In other words, an estimation of a population density of zero must always be accompanied by an estimation of error and confidence level.

Given the inherent imperfections involved in interpretations of no detection, an important gap remains between interpreting the data from an eradication experiment and drawing policy conclusions about the utility of the eradication protocol. Although passage of time to see if the population reappears is the only absolutely reliable method to know if an eradication protocol was successful, even lapses of time are not foolproof; migration into the eradication zone from the outside can always confound the results. Moreover, the longer the time between end of an eradication treatment and judgment of its efficacy, the higher the probability that undetected immigration will occur. Despite the hazards, passage of time is the major accommodation that must be allowed for the inherent limitations in detecting the absence of an "eradicated" target. Further research is needed to specify the appropriate time intervals for different species in different regions.

4. *What are the capacities for migration by pest species?* Migration capacity, like detection sensitivity, is a technical problem of utmost importance to eradication efforts. Experiments guided by the TPM strategy have assumed that a barrier zone would exist between the area freed from the pest and still-infested regions or that the pest could be eradicated again if immigration occurred.

At the moment, knowledge is far too incomplete and unreliable about migratory powers of insects (Stinner et al. 1983). This uncertainty lowers our confidence in our abilities to keep the target pest out even if it is successfully eradicated. Migration studies must be aimed at the abilities of insects to travel under their own powers, in air or water, through the efforts of other organisms, and with the aid of human transportation activities. In addition, more research is needed on the components of a successful quarantine program. Only when migration capacities are well understood can we have any confidence of maintaining an area free of the eradicated pest.

5. *What principles govern the general utility of suppression techniques that have been demonstrated on particular species?* As a tool in eradication efforts, the sterile-male technique is theoretically strong. It has proven, practical effectiveness against the screwworm fly, the Mediterranean fruit fly, and other species (Knipling 1979, 345–62). But we lack the knowledge to help us predict what species are best suited to control by the sterile-male technique, and we need an understanding of the physiological factors that render certain species easily susceptible to the technique. Autocidal methods have proven troublesome, for example, against Colepterans such as the boll weevil and Lepidopterans such as the pink bollworm.

Similar questions surround the use of such other control techniques as classical biological control, augmentation of natural enemies by release programs, pheromones and kairomones, habitat sanitation, and the myriad types of chemicals. In these cases we lack sufficient understanding of the comparative physiology and ecology of the many different insect species. For many known successful suppression techniques, the mechanisms of harm to the target organism are, at best, poorly understood. Applied entomology, despite its enormous growth in the past one hundred years, has barely scratched the surface of the enormous biochemical, ecological, and behavioral diversity within the class Insecta.

Research policy in applied entomology must confront the vast ignorance with which we face our insect competitors. Given the enormousness of the tasks, individual entomologists have understandably relied on hunches and intuition in following up successful research projects. To expect completely rational planning of all entomological research was and still is impossible. Under such circumstances, partisan fervor can easily arise when a scientist is called upon to defend a particular research avenue. Many entomologists believe that a great deal of unsavory politics have been behind some eradication efforts. Proponents of eradication efforts, on the other hand, are equally likely to believe that their critics lack the courage to explore bold advances in technology. Both groups must recognize their shared predicament as partisans dealing with massive ignorance about the complexity of the insect world.

Crucial decisions about whether to launch or halt specific eradication efforts stand in the balance. Initiation of an eradication endeavor is costly, and the resources employed may have better outlets. Conversely, failure to launch an eradication effort could, in some cases, result in more economic and environmental losses than would occur if the project were launched.

6. *What types of biological knowledge are most needed for sound methods of pest control?* Economic entomologists are applied biologists, which means they systematically gather and organize all sorts of information about the biology of insects of economic importance. Biology in this context includes taxonomy, evolution, natural history, population ecology, physiology, toxicology, biochemistry, nutrition, and other subfields of biology. Quick answers are frequently needed in applied entomology, and research funds are always limited. Accordingly, each economic entomologist must make decisions about what sorts of investigations to pursue. Compromises are the normal course of events because of limited research resources.

Proponents of eradication have differed from their colleagues working within the IPM and chemical-control strategies in where they have put their efforts. For example, research on the sterile-male technique was a major investment during the 1960s and 1970s (Hoffmann 1970) for some entomologists because of the technique's utility in eradication. Great advances were therefore made in the fields of insect nutrition and culture, mass-rearing and sterilization methods, and testing for efficacy of the released sterile-males. In contrast, researchers working with the IPM strategies were less enthusiastic about these efforts because they felt the sterile-male techniques would not be useful. IPM envisions efforts to keep an insect below an economic threshold, which is usually higher than the population level at which the sterile-male technique is feasible.

Conversely, entomologists working within the IPM strategy invested great efforts in complex modeling based on time-varying life tables. Such models were useful in exploring interactions between pest-host natural enemies. Eradication efforts against screwworm fly, boll weevil, and other species were, in contrast, based on highly simplified models. These models were admittedly inaccurate, but their utility was sufficient for their purposes, which was to calculate the levels of sterile males needed for release.

No simple algorithm will ever be able to indicate which biological research avenues should have highest priority in our time. Hunches will continue to play a role in making policy on resource allocations in entomological research. Nevertheless, entomologists can do more to compare and contrast their strategic concepts. Through careful studies of the successes and failures of each strategy, we will have the best information possible upon which to base resource allocation decisions.

7. *By what means shall costs, risks, and benefits be measured for insect control technologies?* This final question ties the work of economic entomologists as natural scientists to their roles as social scientists and engineers in the design of practical, acceptable insect-control technology. The answer is addressed in the cost-benefit analysis covered extensively in chapter 3 by LeVeen and will not be discussed further here except to note the importance of considering factors beyond monetary costs and benefits. Assessment of social impact and technology with a broad range of variables is essential in the analysis of eradication proposals, as insect-control technologies affect and are affected by the broad cultural values of society.

Social Issues in Eradication Efforts

The cultural factors that insect-control research and practice affect usually appear as the assumptions, beliefs, and values held by entomologists, agriculturalists, and members of the general public. Sometimes it is easy to recognize and articulate the exact nature of these relevant cultural values. For example, the Federal Insecticide, Fungicide, and Rodenticide Act as amended in 1972 made clear that U.S. society valued both a clean environment and the protection of the chemical industry's proprietary rights. In other cases, however, much work and thought is needed to reveal the deeply hidden values that affect behavior and social standards. Eradication as a concept of insect-control technology has particularly significant links to economic, political, social, and philosophical factors.

Economic and Political Factors

Economic entomology as a field recognizes the role of economic efficiency in the design of insect-control practices. Cost-benefit analyses are generally directed toward efficiency and practicality for the user of an insect-control protocol. Nevertheless, far more than efficiency for the individual farmer is meant by the term *economic*. As noted previously, questions of social equity, fairness, and justice are important dimensions of the economics of a proposed eradication scheme and compose a "political economy" of eradication (and all other modes of insect-control technology). The notion of political economy encompasses the intertwining of economic factors dealing with allocation of resources and political factors dealing with power relationships among people and between individuals and the state.

Benefits from a new technology are generally not distributed equally, either because the invention is not needed by some potential users or because it is inaccessible owing to constraints of money or knowledge. Use of the invention by the early adopters generally returns high profits; clients who adopt later, or never, generally receive few or no tangible benefits (Peterson and Hayami 1977). Proponents of eradication technology have generally resisted the concept of differential benefits from differential adoption rates. Advocates have tended to assume that the projected benefits of eradication are uniformly and benignly distributed to all agricultural producers and to the society as a whole. In at least one case, however, opposition to an eradication scheme developed because of the realization that producers suffering little or infrequent damage from the proposed target would probably derive relatively little benefit compared to their peers who regularly suffered heavier damages (Taylor and Lacewell 1977). In the event of a successful eradication program, the producers who were lightly affected would very possibly have been economically harmed. Producers in the zone freed of the pest would be more efficient competitors, and growers in the lightly-infested area would be faced with higher supplies and lower prices in the marketplace.

Eradication of a serious pest can thus pose direct economic harm to a producer who was affected lightly or not at all by the pest. Eradication, if successful, can effect a redistribution of wealth among growers. Entomologists need to be aware that their creative efforts can lead to a reallocation of income among their clients.

A second feature of the political economy of eradication is that the programs alter the dominant assumptions relating insects, citizens, and state in the United States. Problems with insects have generally been considered a private matter. Founding of the U.S. Entomological Commission, the U.S. Department of Agriculture, and the land grant universities with their experiment stations altered this assumption in the late nineteenth century. Now it is accepted that the government will intervene in the battle between insects and humans under two circumstances: if the insect damage is particularly great or the insect-control technology is particularly hazardous to people or wildlife.

Despite acceptance of government involvement in some aspects of insect control, most citizens feel that these matters should routinely be private, not public, affairs. Chemical control and IPM reflect the predominant view that government should confine its role to research and regulation and not participate as an active implementer. Eradication programs, in contrast, demand a heavy government role in insect-control practices. There is no other way to marshal the personnel and materials, plus the state police power to compel compliance, all of which are essential to eradication projects.

The wisdom of government involvement in eradication programs is debatable. On one hand, a public good is returned if an eradication project relieves a situation in which the pest or chemicals directed at the pest were leading to serious economic loss or to significant environmental damage. On the other hand, entomologists need to consider seriously a number of questions associated with any government's role in eradication work.

How should government handle the problem of unequal access to the benefits of successful eradications, especially for growers suffering a diminished economic return in their businesses? To what extent is the police power of the state a legitimate tool to use against people failing to follow the prescribed manner for an eradication project? Put another way, how moral is it to fine or jail a person for irresponsibility in handling an insect problem? To what extent is it legitimate to treat an area owned or occupied by a person who objects to the treatment? The outrage of urban Californians against the helicopters that dispensed malathion bait in the Mediterranean fruit fly battle illustrates the importance of this question (Marshall 1981; see chapter 6, this volume).

What does one do about the potential for the "insect terrorist," a person who would deliberately sabotage an eradication effort by introducing a pest in violation of quarantine restrictions? The law may be quite clear that such actions will be treated with prescribed punishment, but the law also must enjoy the political support of the citizenry if it is to be effective. Governments must be cautious, as the level of tolerance for penalties for "insect offenses" may be low. Entomologists designing technology dependent on the police power of the state should be acutely aware of the potential for political backlash against laws related to such offenses.

Social Factors

Agricultural producers are a society of atomistic, competitive entrepreneurs (Headley 1980). During the past fifty years, the number of farmers in the United States has fallen from over six million to fewer than three million. Fewer than 10 percent of the farms now produce greater than 50 percent of all farm products marketed. Seen comprehensively, farmers have gone from a vast majority of the American population in colonial times to a tiny minority at the present time (Fite 1981). Those who have stayed in business have done so by managing capital well and by adopting the production technologies best suited to their particular circumstances. Constant changes in technology have been the driving force in American agriculture, and selecting new technology has been the farmer's critical competitive tool. Technologies that were cost-efficient for individual farmers on a short-term basis provided the wherewithal for successful farmers to stay in business.

Chemical control strategies fit admirably into a farming system in which each individual farmer attempted to adopt the most effective technology for his particular circumstances. Chemicals could be purchased and used when needed without cooperation among neighboring farmers. Insecticides that worked were perfect tools for a competitive farmer. In a real sense, chemicals fit the social context of American farming better than either TPM or IPM strategies.

Most IPM work aimed to design technology that could be used on a field-by-field basis. Nevertheless, IPM relied on the importance of natural enemies for insect control. Either natural enemies had to be introduced (classical biological control), augmented, or conserved. Actions between neighboring fields on the same farm or between neighboring farms could affect the populations of natural enemies. Harvesting patterns, drift of insecticides, and varietal selections were all examples of activities in which one farmer could be affected by a neighbor. Effective use of IPM therefore may presuppose an enhanced ability of neighbors to collaborate.

All TPM-inspired programs, but especially those aimed at eradication, require cooperation between neighbors. Effective attack against the total population of an insect necessitates that every pocket of the population be identified and subjected to treatment. Eradication efforts will surely fail if a pocket is left untreated.

To the extent that eradication programs require coordinated production activities on neighboring farms, they go against the social customs that have dominated American agriculture for more than half a century. Farmers have the capacity to collaborate closely when the returns are high enough, but obtaining that collaboration is difficult. Education efforts must be intense, and voluntary cooperation in eradication is unlikely to be sufficient. When the cooperation is obtained for an eradication effort, it is in spite of, not because of, the social organization of farming. In this sense, a high social cost accompanies every eradication effort.

Philosophical Factors

Beliefs about the natural world and the role humans play in it constitute the most difficult, complex, and important of the social factors that affect the work of entomologists. The goal of eradication programs presumes that humans both can and should exercise a controlling influence over the process of natural events. I have previously described the foundations of this belief as "humanistic," because the presumption of human dominance makes humans the dominant force in the natural world (Perkins 1982). Proponents of eradication, particularly those working within the TPM strategy, were optimistic that, given enough research, technology could be developed to manipulate agroecosystems in powerful ways, including the eradication of well-established species. The farming industries based upon capital-intensive methods of agriculture and the chemical-control strategy are also humanistic in their metaphysical basis. Eradication proposals, capital-intensive farming, and chemical control therefore share important characteristics in the mind of the contemporary farmer. To the extent that all of these strategies produce cost-effective technologies they gain easy access into the repertoire of the knowledgeable farmer.

In contrast, IPM was developed by scientists who believed it was impossible or unlikely that humans could successfully and completely manipulate the natural world. I have previously characterized IPM's assumptions as "naturalistic," because the strategy presumes a predominance of natural forces over human technology (Perkins 1982). Consider the words of Robert van den Bosch (1978), one of the foremost architects of the IPM strategy: "Our problem is that we are too smart for our own good, and for that matter, the good of the biosphere. The basic problem is that our brain enables us to evaluate, plan, and execute. Thus, while all other creatures are programmed by nature and subject to her whims, we have our own gray computer to motivate, for good or evil, our chemical engine. Indeed, matters have progressed to the point where we attempt to operate independently of nature, challenging her dominance of the biosphere. This is a game we simply cannot win, and in trying we have set in train a series of events that have brought increasing chaos to the planet."

Van den Bosch was specifically criticizing the use and abuse of insecticides, but IPM, his alternative to reliance on chemical control, incorporated an assumption that humans must learn to rely on natural processes, such as natural enemies and host-plant resistance. Moreover, total success was not likely in an intervention as strong as eradication. IPM adherents believed, in addition, that eradication was unnecessary on practical grounds.

Beliefs about the relationships of humans to natural forces are not the only point of philosophical division between proponents of eradication and advocates of other types of control. Specific moral considerations attend any proposal to eradicate an insect species. Most eradication schemes have not aimed at global or complete annihilation of a species, but successful eradication in one place leads inexorably to the theoretical prospect of deliberate extinction of a pest. The moral question at issue is whether the targeted species has any intrinsic rights to survive.

Judgments on this matter have not been thoroughly formulated, and the variations expressed to date are not well correlated with adherence to the TPM or IPM strategies. Opinions among entomologists include (a) considering the question absurd, that is, assuming that the target of eradication has no rights, (b) asserting that extinction is a natural process and therefore moral questions surrounding eradication are moot, and (c) believing that humans have no right to annihilate that which cannot be created, essentially an assertion that global eradication of a species would be morally wrong.

It is beyond the scope of this chapter to engage in a thorough discussion of the ethical aspects of eradication. The moral concerns taken together with the different viewpoints regarding the relationship of human beings to nature indicate that important questions attend the proposed eradication of any species. Unexamined philosophical differences between peoples can generate extreme passions that may potentially destroy all constructive efforts to control an insect population.

Important questions and issues of a scientific and social nature surround every proposal for the eradication of a pest species, even an eradication that is not global in scope. Proponents and opponents of eradication exercises must allow open debate of these issues with a sense of self-criticism if economic entomology is to make significant progress toward resolving the concerns. It is especially important to recognize that these dilemmas cannot be resolved simply by so-called objective decisions made unambiguously from relying on scientific experimentation only. Instead, the choices to be made require an examination of empirical data plus an examination of values, beliefs, and assumptions, many of which are not scientific in nature. As I have stated previously, "Insects may be small and invite contempt, but efforts to deal with them evoke all the most deeply held beliefs about what it is to be human" (Perkins 1982).

Literature Cited

Baker, H. 1952. Spider mites, insects and DDT. In *Insects: The yearbook of agriculture*. Washington, D. C.: Government Printing Office.
Baumhover, A. H., A. J. Graham, B. A. Bitter, D. E. Hopkins, W. D. New, F. H.

Dudley, and R. C. Bushland. 1955. Screwworm control through release of sterile flies. *J. Econ. Entomol.* 48:462–66.

Brown, A. W. A. 1961. The challenge of insecticide resistance. *Bull. Entomol. Soc. Amer.* 7:6–19.

Bureau of the Census. 1957. *Census of manufacturers, 1954.* Vol. 2, pt. 1, SIC 2897. Washington, D. C.: Government Printing Office.

Carson, R. 1962. *Silent spring.* Boston: Houghton, Mifflin.

Dunlap, T. R. 1980. Farmers, scientists, and insects. *Agric. History* 54:93–107.

Economic Research Service. 1969. DDT used in farm production. *Agricultural economic report* 188. Washington, D. C.: USDA.

Fite, G. C. 1981. *American farmers: The new minority.* Bloomington: Indiana Univ. Press.

Forbush, E. H., and C. H. Fernald. 1896. *The gypsy moth.* Boston: Wright and Potter.

Graham, H. M. 1978. Sterile pink bollworm: Field releases for population suppression. *J. Econ. Entomol.* 71:233–35.

Hagen, K. S., and R. F. Smith. 1947. First annual report of the entomologist for the Westside Alfalfa Pest Control Association. Mimeo.

Headley, J. C. 1980. The economic milieu of pest control: Have past priorities changed? In *Pest control: Cultural and environmental aspects,* ed. D. Pimentel and J. H. Perkins, 81–97. AAAS Selected Symposium 43. Boulder, Colo.: Westview.

Hearn, A. B., P. M. Room, N. J. Thomson, and L. Y. Wilson. 1981. Computer-based cotton pest management in Australia. *Field Crops Res.* 4:321–32.

Hoffmann, C. H. 1970. Alternatives to conventional insecticides for the control of insect pests. *Agr. Chem.* 25(9):14–19a; 25(10):19, 21–23, 35.

Ives, P. M., L. T. Wilson, P. O. Cull, W. A. Palmer, C. Haywood, N. J. Thomson, A. B. Hearn, and A. G. L. Wilson. 1984. Field use of SIRATAC: an Australian computer-based pest management system for cotton. *Prot. Ecol.* 6(1):1–22.

Knipling, E. F. 1953. The greater hazard: insects or insecticides. *J. Econ. Entomol.* 46:1–7.

———. 1966. Some basic principles in insect population suppression. *Bull. Entomol. Soc. Amer.* 12:7–15.

———. 1979. *The basic principles of insect population suppression and management.* USDA handbook 512.

Lyle, C. 1947. Achievements and possibilities in pest eradication. *J. Econ. Entomol.* 40:1–8.

Marshall, E. 1981. Man versus medfly: some tactical blunders. *Science* 213:417–18.

Metcalf, C. L., and W. P. Flint. 1939. *Destructive and useful insects.* 2d ed. New York: McGraw-Hill.

Metcalf, C. L., W. P. Flint, and R. L. Metcalf. 1951. *Destructive and useful insects.* 3d ed. New York: McGraw-Hill.

Michelbacher, A. E., and O. G. Bacon. 1952. Walnut insect and spider-mite control in northern California. *J. Econ. Entomol.* 45:1020–27.

Michelbacher, A. E., and R. F. Smith. 1943. Some natural factors limiting the abundance of the alfalfa butterfly. *Hilgardia* 15:369–97.

National Academy of Sciences. 1975. *Pest control: An assessment of present and alternative technologies.* Vol. 3, *Cotton pest control.* Washington, D. C.: National Academy of Sciences.

Newsom, L. D. 1974. Pest management: history, current status and future progress. In *Proceedings of the summer institute on biological control of plant insects and diseases,* In ed. F. G. Maxwell and F. A. Harris, 1–18. Jackson: University Press of Mississippi.

Office of Technology Assessment. 1979. *Pest management strategies in crop protection.* Vol. 1. Washington, D. C.: Government Printing Office.

Perkins, J. H. 1978a. Edward Fred Knipling's sterile-male technique for control of the screwworm fly. *Environ. Rev.* 5:19–37.

———. 1978b. Reshaping technology in wartime: the effect of military goals on entomological research and insect-control practices. *Tech. Cult.* 19:169–86.

———. 1980. Boll weevil eradication. *Science* 207:1044–50.

———. 1982. Insects, experts, and the insecticide crisis. New York: Plenum.

Peterson, W., and Y. Hayami. 1977. Technical change in agriculture. In *A survey of agricultural economics literature,* vol. 1, ed. R. Martin, 528–32. Minneapolis: University of Minnesota Press.

Proverbs, M. D., J. R. Newton, and D. M. Logan. 1977. Codling moth control by the sterility method in twenty-one British Columbia orchards. *J. Econ. Entomol.* 70:667–71.

Rohwer, S. A. 1945. Report of the special committee on DDT. *J. Econ. Entomol.* 38:144.

Scruggs, C. G. 1975. *The peaceful atom and the deadly fly.* Austin: Pemberton.

Smith, R. F., and R. van den Bosch. 1967. Integrated control. In *Pest control,* ed. W. W. Kilgore and R. L. Doutt, 295–340. New York: Academic Press.

Stern, V. M., R. F. Smith, R. van den Bosch, and K. S. Hagen. 1959. The integrated control concept. *Hilgardia* 29:81–101.

Stinner, R. E., C. S. Barfield, J. L. Stimac, and L. Dohse. 1983. Dispersal and movement of insect pests. *Ann. Rev. Entomol.* 28:319–35.

Taylor, C. R., and R. D. Lacewell. 1977. Boll weevil control strategies: Regional costs and benefits. *So. J. Agric. Econ.* 9(1):125–35.

U.S. Congress. 1952. House. *Investigation of the Use of Chemicals in Foods and Cosmetics,* Report 2356, 82:2, part 1.

———. 1965. Senate. Committee on Appropriations. *Agricultural appropriations for 1966: Hearings,* part 1. Washington, D. C.: Government Printing Office.

———. 1966. Senate. Committee on Appropriations. *Agricultural Appropriations for Fiscal Year 1967: Hearings,* part 1. Washington, D. C.: Government Printing Office.

van den Bosch, R. 1978. *The pesticide conspiracy.* New York: Doubleday.

Welch, S. M. 1984. Developments in computer-based IPM extension delivery systems. *Ann. Rev. Entomol.* 29:359–81.

Williams, T. H. 1970. *Huey Long.* New York: Alfred A. Knopf.

Whalon, M. E., and B. A. Croft. 1984. Apple IPM implementation in North America. *Ann. Rev. Entomol.* 29:435–70.

Whorton, J. C. 1974. *Before silent spring.* Princeton: Princeton Univ. Press.

Wilson, L. T., R. E. Plant, T. H. Kerby, L. Zelinsky, and P. B. Goodell. 1987. Transition from a strategic to a tactical crop and pest management model: Use as an economic decision aid. *1987 Proceedings Beltwide Cotton Production Research Conferences:* 207–13.

3

Economic Evaluation
of Eradication Programs

E. Phillip LeVeen

This chapter will discuss the difficulty of valid economic evaluation of any proposed eradication policy. Each eradication program is unique and has no history—therefore we do not have and cannot obtain the information required for such evaluation. Because the effectiveness of eradication programs depends mainly on toxic chemicals, these programs are inherently risky, posing potential long-term threats that are not easily quantified or anticipated. Human inability to adequately assess the effectiveness of such programs or their long-term consequences suggests that a far safer and potentially more effective approach would be the development of programs that would manage unwanted pest populations through nonchemical means at some nonzero level.

Two general kinds of disputes can arise in economic evaluation: those relating to the accuracy of the basic empirical analysis and those relating to the acceptability of the values that underlie the economist's judgment of a policy's desirability. All economic evaluations consist of two classes of statements. "Positive" statements are empirical claims about the way any particular policy will alter the world. An example of such a statement is: "the elimination of the Medfly will increase the production of oranges by x percent." Underlying such a statement is supposed to be a set of empirically derived relationships that could, in theory, be subjected to verification through independent experiments. I call this statement "positive" because, in principle, it can be empirically verified or rejected.

"Normative" statements are also a necessary part of economic policy evaluations; such statements are based on an underlying philosophy and a related set of values that allow the economist to judge whether a policy is "good" and thus whether it ought to be pursued. Unlike positive statements, normative claims are not testable by any commonly accepted scientific criteria. For example, the statement "the elimination of the Medfly is good policy because it will increase food production by x percent" is more than simply an assertion about the outcome of the policy. The idea that the policy is desirable because it increases output (granting, for the moment, that eradication does increase output) involves the value that more is better than less. Although such a value enjoys wide acceptance in our society, there are many who may not accept it and therefore will not accept the economist's prescription, no matter how accurate the positive analysis.

It might be supposed that disputes over positive statements can be readily resolved, since positive statements are susceptible to scientific verification. In reality, however, when one lacks the data required for such empirical verification, an element of uncertainty cannot be eliminated. Two evaluations may thus lead to very different conclusions, if critical areas of uncertainty are treated differently. Unfortunately, the level of uncertainty in evaluation of eradication programs is large and probably irreducible.

Debates over values can be productive, though there is no simple criteria for judging one set of values as superior to another. It may be useful to separate disputes of fact and value to reduce the scope of debate, but such separation will not eliminate debate itself. Some might suggest that we stick to the "facts" and leave the values out, yet ultimately we are striving to understand what is good policy, and inherent in the idea of "good" are values. Even if we try simply to analyze the facts we must exercise some judgment since not all of the facts are known; we must make assumptions, and this act involves making value judgments. In short, there is no truly "value-free" approach; should not the debate openly confront the issue of values rather than pretend that important social choices can be reduced to scientifically verifiable propositions?

The following analysis illustrates the kinds of obstacles one encounters in any evaluation of an eradication program. As will become apparent, these obstacles cannot now be overcome, given our insufficient information base and the lack of consensus over certain important critical values.

The Concept of Economic Evaluation

Let us suppose that a public agency is deciding whether or not to undertake an eradication program. How, then, in an ideal world might an economic evaluation assist in this process? We will assume that the evaluator has access to complete and accurate information on any issue relating to the evaluation program; there is no uncertainty. What information would the evaluator require?

The basis of such an evaluation is a detailed specification of how the proposed eradication plan would change the world relative to what would occur without the policy (or in comparison to some alternative policy). Such a specification requires several important kinds of information. First, the entomologist must be able to describe what will happen if the pest is not controlled by a publicly-supported eradication plan. Such a prediction, projection of the future pest population, requires an understanding of the life-cycle of the pest and its interactions with other pests and with other components of the relevant ecosystem.

Having identified the population levels that will prevail without eradication, the evaluator must quantify their impact on society. From entomologists, agronomists, medical researchers, economists, and sociologists, the evaluator would seek to learn the extent to which the uncontrolled pest presents a threat to human or animal health, the food supply, property, or any other relevant social concern. Defining these impacts requires a further analysis of how society will respond to the future pest populations. For example, farmers may adapt their own pest control efforts to reduce the adverse impacts of the pest, thus mitigating the initial threat to the food system; these reactions and their consequences must also be incorporated into the evaluation of the world picture *without* the public eradication program.

In order to assess the impact of the eradication policy, the evaluator must next assess how the specific policy will influence the pest population and how, in turn, the altered population will affect human health, food supply, and property. To assess the full impacts of the eradication policy, he or she also must know the indirect consequences of undertaking the policy. For example, if the public sector devotes significant resources to the program, does this mean some other program will be cut back? Or will the resources come from the private sector through additional taxes, meaning that the private sector will curtail some other activities? These possibilities must be considered part of the overall decision to undertake eradication.

Finally, the eradication program itself may have impacts other than those on the targeted pest population. The program may affect other pest populations, both in the short and long term, which may then affect health, food supply, and property. Similarly, the program itself may have a direct impact on human health, food supply, and property: Individuals exposed to eradication chemicals could experience increased health problems; chemicals could become part of the food chain or could directly damage property.

By comparing all of these aspects with and without the eradication program, the evaluator can describe the consequences of the eradication policy. But impacts are described in terms of such diverse and noncomparable measures as incidence of health problems, loss of human productivity, reduction in food productivity, property losses, and disruptions to the ecosystem. No matter how accurately described these impacts may be, such facts by themselves may not determine if the policy is desirable since in most cases there will be desirable and undesirable impacts associated with each policy option. The problem is how to compare these options for which the impacts are not easily compared.

To permit the comparison of diverse impacts, economists use a tool known as cost-benefit analysis. By arraying the positive impacts against the negative ones and by converting all impacts to a common measure—dollars—this tool allows calculation of the overall impact of a policy on society's income. If the dollar measure of the positive impacts is greater than the dollar estimate of negative impacts, then the policy increases the overall income of society and is said to be in the public interest and therefore worth pursuing. If the costs outweigh the benefits, then the eradication policy would make society worse off than a policy of noncontrol.

Although cost-benefit analysis appears to answer the difficult question of how to compare policy options and choose a particular policy, its success depends on whether dollars and income adequately express the full range of social values involved in such a decision, even granting that a dollar amount can be given to impacts that may not have commonly recognized economic values. Before entertaining a more detailed discussion of the issue of values, however, let us first consider the "real world" problems of evaluation with the ideal version as described here.

Real World Problems: Lack of Basic Scientific Data

The idealized model of evaluation above assumed complete and accurate information on all relevant aspects of the policy issue. The real world is characterized by poor information and high degrees of uncertainty. Eradication policies are particularly poor candidates for rational evaluation, as many of the basic facts necessary to their evaluation are simply unknown.

Eradication policies are usually aimed at a pest that is in the process of becoming established in a new environment. Because there is no history, it is impossible to assess with certainty whether the pest can adapt and become a permanent resident of the region. If it can establish itself, how will the pest interact with the rest of the local ecosystem? How will local environmental factors influence its population if there are no controls? Answers to these questions are not available from direct observation, and if enough time passes to allow the pest to become established (and hence provide the basic data needed to answer them) eradication programs might not be effective. Such policies are predicated on eliminating the pest *before* it can become established.

There may be situations in which the pest has successfully established itself in other, similar regions that can serve as a base for projecting the probable impacts of the pest on the new region. This assumes that the pest has been adequately studied in other regions. But because no two regions of the world are identical in all respects, even if there is information available, it may not provide a good indication of how the pest will adapt to the new region. In many cases, the basic research necessary for the evaluation has not been undertaken. At best, the evaluator must depend on indirect and relatively speculative evidence, and, at worst, he or she has almost no scientific basis for estimating how large the uncontrolled pest population is likely to be if eradication is not undertaken.

Without the empirical base needed to establish the likely magnitude of the pest population it is impossible to derive scientifically valid estimates of the pest's probable impacts should it establish itself. In the absence of such information, the evaluator must choose from widely differing speculative claims. No matter which set of claims is chosen, the analysis will be subject to criticism from the opposing camp. Such debate cannot be resolved unless the basic research is available.

The evaluator's problems are compounded by other unresolvable questions. In the ideal case, the evaluator must compare the world with and without the eradication policy. But an essential issue here is whether eradication is really possible.

"Eradication" is an appealing concept because it promises a once-and-for-all solution to the potential pest problem. Presumably, eradication is preferred to an ongoing strategy of managing the established pest population; it is simpler and promises total elimination of all pest-related problems. But if eradication is technically impossible, future damages may not be eliminated by the program, and eradication involves a long-term strategy of population management that attempts to maintain a zero-level pest population by repeated application of the controls.

Thus, in considering the possible impacts of the eradication program, the evaluator must know if the measures will have to be repeated. If it is clear that eradication may not be possible or that its success is subject to considerable uncertainty, then much of the urgency usually associated with implementing these programs may be unnecessary. There may be more time for research and study of the pest to allow the development of a long-term population control program. In all likelihood, a long-term management strategy would differ from a series of "eradication" programs intended to maintain a zero population of the unwanted pest.

Because of the lack of scientific research about the pest, the potential effectiveness of a proposed eradication program may be uncertain or at least a matter of considerable debate. There are examples of successful eradication programs and others where "eradication" appears to be almost an annual program. Here the evaluator must again decide on which of the competing claims to base the analysis.

In addition to the question of effectiveness, the proposed eradication program also raises questions regarding its possible adverse effects. Virtually all eradication programs are based on the use of toxic chemicals that offer immediate results. Toxic chemicals pose possible threats to the health of humans and animals, and the full implications of this exposure may be unknown. No matter how carefully implemented the program, accidents leading to widespread exposure are a risk inherent in the use of these chemicals.

Exposure to toxic chemicals may result in acute illness or death; it also may result in chronic health problems that do not appear for many years. Although a chemical may be described as safe, it can actually carry a long-term impact that has yet to be scientifically observed. Many chemicals once used extensively and believed harmless to human health (for example, Agent Orange) have been subsequently found to be dangerous as the long-term effects have become apparent.

Chemicals used in eradication programs, usually characterized as broad spectrum pesticides, may have adverse impacts on nontargeted pests. Some of these pests may serve useful functions. Disruption of these beneficial functions may cause damage to health, property, or the food supply. Additional interventions may then be required because the controls of natural biological processes have been altered.

The chemicals themselves may also pose a threat to property—for example, the malathion used to eradicate the Medfly damaged enamel car finishes (see chapter 6, this volume). These possible adverse impacts should also be incorporated into any policy analysis.

The difficulty in estimating the impacts associated with eradication chemicals is apparent, especially since many of the possible long-term health effects from exposure to chemicals take years to develop. Because human health is influenced by diverse factors and because knowledge regarding many diseases is incomplete, it is difficult to prove specific cause-and-effect relationships in a scientifically acceptable manner. Similarly, our knowledge of complex ecological relationships is far from sufficient to permit a careful assessment of the consequences of using the chemicals.

Problems in Positive Economic Evaluation

Once the evaluator has decided how the eradication program is going to affect biological systems, there remains the important task of projecting how eradication is going to influence social and economic systems. If the program is aimed at a pest that threatens agricultural output, for example, it will be necessary to calculate the farm income that will be saved and any improvement to consumers who may experience lower food costs if the pest is eradicated. Similarly, since the program will require the expenditure of public revenues, it will be necessary to calculate what the program costs, in terms of either the actual resources used or the other opportunities that must be forgone to fund this program.

The quality of the socioeconomic analysis depends entirely on the quality of the basic analysis of biological impact. No matter how well constructed the economic analysis, if it is based on faulty assumptions about the effectiveness of the eradication program, its results will be worthless. Nevertheless, there are problems specific to the positive socioeconomic analysis that deserve comment.

A very common error made in economic evaluations of pest-control strategies is the failure to look at the overall effects on the entire economic system; instead there is a tendency to concentrate on the immediate effects on affected producers. One cannot infer the total impact of a policy from summing the adverse individual impacts. For example, in economic evaluations of the possible effects of banning the use of the nematicide DBCP (1,2–dibromo–3–chloro–propane) in United States agriculture the USDA found that the ban would reduce the output of citrus, soybeans, peaches, and other crops attacked by nematodes. It was then concluded that the smaller output would result in large losses of revenue to growers of these crops. The loss in revenue was used as a measure of the cost of the ban. Such an analysis is misleading for several reasons.

The claim that total farmer revenues would be reduced because of losses in output would be true only for an individual producer whose output was adversely affected by the ban. Suppose the ban affects all producers equally and causes everyone to lose ten percent of a crop; it does not follow that revenues will be lower by the same ten percent. To anticipate what a ten percent loss in output will do to revenue, one must estimate how the loss will influence the commodity's price. According to a well-established principle of economics, the reduction in supply will cause the price of the commodity to rise. How much it will rise depends on the price elasticity of the demand for the commodity. In virtually all agricultural commodities, this price elasticity is less than "one," which means that for every one percent reduction in supply the price will rise by more than one percent. This being the case, the loss in agricultural output would actually increase producer revenues, not decrease them. The ban would thus actually benefit producers rather than injure them, in much the same way as many other agricultural policies deliberately attempt to restrict output so as to increase farm income.

One might ask why producers would fight the ban if it would benefit them. Part of the answer is that producers (along with some USDA economists) may be unaware of this beneficial relationship. Another factor is that the ban's effect is not equally distributed over all producers. Many producers do not have nematode problems, so the ban on DBCP would have no impact on their output. Other producers may lose a significant part of their crop and be forced to grow other less profitable crops as a result. In this case, the losses of the latter group will cause a general increase in price that benefits those who have no nematode problems or lost output. The price rise may not be sufficient to fully compensate those with nematode problems for losses occurring because of the ban. For the overall industry, income will be higher with the ban, but there will also be individuals who suffer losses. Evaluations that concentrate on the losses without also estimating the gains present an incomplete and misleading economic accounting. Nevertheless, this is common practice.

In the same vein, many economic evaluations fail to recognize that the underlying economy is dynamic and gradually changing in response to new influences, mitigating the initial impact of a pest outbreak. If analysis concentrates on short-term, immediate impacts of the pest and neglects the adaptive responses taking place, the conclusions will overemphasize the potential gains of eradication.

To illustrate these adaptive effects, let us refer to another controversy—the use of Compound 1080, a chemical used to exterminate coyotes. Compound 1080 was banned in 1972. In 1982, the EPA held hearings concerning the impacts of reregistering this chemical for use on coyotes, animals said to be increasingly a problem for the sheep industry. In the debate over 1080, those favoring use of the chemical argued that under the ban more sheep and lambs were being killed. This led to smaller supplies of meat and wool, thus causing lower producer incomes and higher consumer prices. If one accepts the argument that restrictions on the use of 1080 indeed have increased sheep losses, a point that was strongly debated, it does not follow that producers and consumers are worse off. The price-elasticity effect discussed above tended to reduce the impact of the ban.

Producers have other options; they can shift to other methods of controlling coyote predation. These alternatives may not be as effective or as cheap, but, if the value of the sheep protected exceed the costs of the alternative control techniques, growers will have incentives to use these alternatives. Should the alternatives prove too costly or not sufficiently effective, growers can shift their operations to other kinds of livestock, such as cattle, which are not as vulnerable to coyotes. The net effect of these changes will be to mitigate the loss of the chemical as a control strategy.

Not all sheep growers have predation problems; flocks kept under close scrutiny within secured pastures have fewer losses than flocks on open ranges, especially on public land (where sheep growers can obtain very cheap grazing permits). If predation rates rise for these latter growers, reducing overall supplies of sheep and wool, the resulting higher prices will increase the profits of those without predation problems. The increased profits of this group would encourage greater investment in flocks located in predation-free zones.

The resulting increase in sheep production in predation-free areas would, after some time, mitigate the higher initial prices. It might be added that these higher prices would also encourage foreign producers to sell more wool and lamb to the United States, further reducing the impact of higher prices for U.S. consumers.

Consumers can adjust to higher sheep prices by shifting their purchases to other foods that substitute for the higher cost of lamb. Such changes in consumption might initially drive up the costs of the substitute commodities, but the higher prices would also signal producers to increase output of these substitutes, thus reducing the initial price effects.

If we include these dynamic adjustments of the entire system in our impact analysis, the effect of the ban on 1080 would be substantially smaller than it might at first appear. One needs a holistic or systemwide analysis to capture the full impact of a policy on the nation's income.

Further Problems for Economic Evaluation: The Value Issue

At the outset we distinguished between positive and normative statements; to this point we have concentrated on the problems of analyzing the actual impacts of an eradication policy. Now it is time to examine the value, or normative, issues.

In the ideal evaluation, the decision whether or not to proceed with eradication is based on a comparison of benefits with costs. In order to make such a comparison, it is necessary to quantify the various impacts of the eradication program—that is, to place a monetary value on them. Yet in the real world it is often extremely difficult to express important effects in terms of dollars, even with accurate estimates of the physical changes that will result from the eradication program. In the course of trying to assign economic values, we encounter most of the important normative issues that cannot be resolved through the acquisition of better data.

Placing a Value on Life, Health, and Other Nonmarket Impacts

The difficulty in measuring impacts in terms of dollars is most pronounced for physical impacts that have no obvious market representation. The impacts of illness and death have no evident market value. Economists have invented many proxy values for illness and premature death to allow these impacts to be incorporated into evaluations. For example, illness prevents the individual from working and requires expenditures on medical treatment. Lost work time plus the medical costs can thus be used to represent an increased incidence of acute of chronic illness from exposure to toxic chemicals. Similarly, premature death represents lost productivity that has a market value—that is, the income not earned as a result of early death. Yet it is widely accepted that the value of life is more than simply the dollar amount that an individual produces.

Economists have used awards by juries as another approach to measuring how much society values life. They have also estimated how much individuals are willing to pay to reduce the risk of death, through, for example, the purchase of safety devices on cars. But none of the measures of value devised so far have met with general acceptance. Other than for national defense, few politicians will go on record in favor of a policy that explicitly increases the incidence of premature death, even if this can be justified in terms of large economic benefits.

In short, though economists may have found "rational" measures of the economic value of life, these measures are not generally approved by decision makers who feel that life should not be traded against other material benefits. Of course, policies that implicitly involve such trade-offs are made all the time, but generally the trade-off is not explicitly addressed. Since economic evaluation attempts to make all impacts explicit, the lack of consensus on how to treat the value of life is a major obstacle to the analysis of any program that would expand the use of toxic chemicals.

Other kinds of impacts also raise value issues. For instance, the disruption of the ecosystem after the introduction of a new pest or the use of chemicals intended to control the new pest could have far-reaching implications for plant and animal life. Some of these effects may have direct economic significance that can be measured in terms of changes in food supply and property value, but some of these may not have easily identified economic values.

In the case of the ban on Compound 1080, the analysis revealed that other species of animals would be adversely affected by the reintroduction of this poison. Although most of these animal populations have no immediate economic value, their existence is seen as worthwhile, and many individuals have fought government efforts to reregister the chemical. There is no market whereby concerned individuals can purchase "preservation rights" for wild animals, and there are no obvious proxy values for use in placing an economic value on the benefits of protecting these animals. Yet few would deny that some benefit does exist, and therefore, any evaluation that omits such benefits is missing a potentially important impact of the policy.

Valuing Uncertain and Future Program Impacts

The poisoning of nontarget animal populations raises further important issues in assigning economic value. Disruptions to the ecosystem potentially have long-term, adverse impacts that could profoundly affect future generations. In recent years, for example, there has been a growing awareness of the need to maintain diversity, especially of genetic material for breeding, in order to preserve our capacity for adapting to changes in the future environment. Consequently, the loss of several species may reduce diversity and eventually lead to large, if unknown, economic impacts on the quality of life for future generations.

Most of these possible future economic impacts cannot be carefully defined or assigned an economic value today. At best, we may be capable of assigning a range of possible outcomes along with some probability distribution of each outcome. Assuming this capability, it is then possible to calculate an "expected future value" of the event by multiplying each outcome by the probability of its occurence. In this way, uncertainty can be incorporated into the evaluation. However, it is unlikely that this procedure will have much impact on the choice of strategy; no matter what value is assigned to an uncertain future event, it will have little weight in the analysis because it will be heavily "discounted."

Discounting is employed to insure that present and future economic values are comparable. If one had a choice of two strategies, both requiring the same initial investment and both having the same payoff in terms of dollars, but one produced its payoff in a year and the other in five years, the obvious choice would be the strategy with the earliest payoff. The money received from the earlier payoff could be reinvested for the remaining four years in some other use and could produce more income during that time.

The two strategies are not really equivalent; the benefits of the second strategy, while nominally the same as those of the first, are really smaller by the amount of interest that could be earned over the four year interval. Thus, to compare the payoffs, we must express them in comparable terms; this is accomplished by reducing the value of the later payoff by the interest forgone. This is called discounting. The procedure can be applied to a benefit or a cost.

The greater the interest rate or the further out in time the event occurs, the lower will be its value measured in terms of current dollars. Even at relatively low interest rates, events occuring twenty to thirty years in the future will have almost no value in present value terms. For example, a dollar in thirty years, discounted at 5 percent, has a present value of about 23 cents; a dollar in fifty years is worth 8 cents today. A dollar in thirty years discounted at 7 percent is worth 13 cents today; a dollar in fifty years discounted at 7 percent is worth 3 cents today.

Eradication programs may produce immediate benefits that are greater than direct program costs, but their indirect costs to human health or the stability of the ecosystem may not become apparent for many years. If we employ the techniques described above to evaluate these hidden costs, they may have little consequence for the evaluation.

For example, suppose we estimate that a given program has a 1 percent chance of producing hidden costs that will not become apparent for fifty years, having a value of $1 billion. There is a 99 percent chance of no long-term hidden costs. Using the above technique, we would estimate the expected value of future costs equal to $10 million in fifty years (1 percent of $1 billion plus 99 percent of zero costs), and then we will discount $10 million at, say, 7 percent for fifty years, which gives a present value of about $300,000. This $300,000 cost may be easily outweighed by the immediate program benefits, which are not discounted or reduced by uncertainty.

While there appears to be a strong economic argument for discounting future benefits or costs, it should be clear from this example that discounting sharply reduces the importance of future generations in such evaluations. If the future generations could be present today and make their preferences known, it is unlikely that they would accept a program that increases short-term economic benefits at their long-term expense.

The built-in biases that favor the present over the future and the certain over the uncertain might well be justified in evaluating business investments that have no potentially devastating long-term impacts on the environment or human health, but these biases are inappropriate when evaluating eradication programs that do carry long-term potential threats. The issue raised by the discounting procedure is essentially a moral one—does the present generation have the right to willingly neglect potentially catastrophic and irreversible consequences of its actions on future generations?

Evaluating the Distribution of Benefits and Costs

Thus far nothing has been said about winners and losers from eradication programs and how the distribution of benefits and costs is treated in evaluation. Most cost-benefit analyses simply report an overall ratio of benefits and costs aggregated over all groups without specifying who will actually receive the benefits. It is assumed that if total benefits exceed total costs then society is better off. But what if the benefits accrue to the wealthy and the costs accrue to the poor? Such a regressive distribution is consistent with a positive cost-benefit ratio; economists might thus declare the policy desirable, but many others would find the negative distributional impact significant enough to dismiss the policy.

Economists do not incorporate distributional considerations in their cost-benefit analyses because of a core value underlying neoclassical economic theory. In order to measure the impact of taking money from one person and giving it to another, one must be able to compare the benefits received by the winner and the costs incurred by the loser. Can we say that a dollar given up by an already poor person imposes greater costs on that individual than the additional benefits conferred by the dollar on the already wealthy person? Although most of us might intuitively accept this relationship, a central proposition of economic theory asserts that it is impossible to make comparisons of this kind; there is simply no objective basis for making such measurements.

Economists contend that cost-benefit analysis is solely concerned with "efficiency"—that is, finding the allocation of societal resources that produces the greatest income. Because we cannot measure how different distributions of income influence overall social welfare, it is not possible to incorporate distributional considerations into the analysis of efficiency.

In defense of their sole concern with efficiency, economists further argue that if a policy produces greater benefits than costs and hence is efficient it would be possible, in principle, for the winners to use their additional benefits to fully compensate the losers and still have income left over. Therefore, efficiency and equity can be compatible. That there are few mechanisms for achieving such compensation and that it almost never actually takes place is generally overlooked in this defense of efficiency.

Distributional issues are potentially important in virtually all social policy and are specifically relevant to eradication policy. The trade-off of future for present already discussed is really a distributional issue. Another concern is where to draw the boundaries for the policy evaluation. If one concentrates on a very small region and on only one group within that region, the results will be very different from if one chooses the entire nation and all groups as the relevant impact area. Many studies simply address the impacts on the producer or property owner in a particular region. Insofar as the producer group is part of a much larger system of producers, this narrowing of the focus encourages the neglect of the important systemwide impacts, as described in examples above.

In general, cost-benefit analysis is intended to capture the entire range of impacts on all groups throughout the nation. Even this is somewhat narrow, given the increasingly integrated nature of the world economy. If a pest threatens a particular group of citrus producers in one region of California, should the analysis be restricted only to those with adverse consequences? Obviously those whose income may decline would want the public to be aware of their unique situation, but what if producers elsewhere in California or in some other state could increase their production to offset the regional impacts? Consumers may not be adversely affected, and other producers may be benefitted. From this larger perspective, the threat to the citrus growers does not appear particularly profound.

In short, the issue of where to draw boundaries is essentially a distributional question; one person's loss may be another's gain. There may clearly be some need to redress the loss, even if someone else does benefit, but from the perspective of efficiency there is no way of addressing this issue. At best, an analysis that fully describes all impacts, calling attention to the distributional issues, as well as the overall consequences of a policy, would allow a better political determination of the best course. Unfortunately, few evaluators adopt this approach; instead, they implicitly incorporate a particular distributional perspective by narrowing their scope and do not produce a complete picture; thus they deprive decision makers of the information needed to make an equitable policy choice.

Distributional issues take other forms as well. For instance, in the evaluation of the 1982 gypsy moth eradication program in Santa Barbara, California, the primary benefits were associated with the preservation of shade trees in the suburban environment. It was asserted that the unchecked gypsy moth would eventually destroy shade trees, the loss of which would pose a problem for landscaping homes. Without shade trees, the value of the homes would be lower. Preventing this property loss was the primary economic benefit of the proposed eradication program.

The analysis was carried out in the Santa Barbara area, which has some of the highest housing values in the nation, with median house costs in excess of $300,000. Using a formula that relates housing value and the presence of shade trees, the economic analysis estimated the possible costs of allowing the gypsy moth to establish itself in this region. Since the formula attributes a percentage of housing value to the presence of shade trees, it should not be surprising that the analysis found a potentially huge economic impact if the pest were allowed to proliferate. By using Santa Barbara's housing stock as the basis for the analysis, the percent reduction in value was very much greater than had a low-income housing area been used. Indeed, the study concluded that losses from the gypsy moth in the Santa Barbara area would exceed a billion dollars in this one community alone. Such large losses justified any expense associated with eradication.

The issue raised here is not so much whether this methodology is appropriate, though I have strong reservations regarding the use of such a formula. What is of greater concern is the use of a very affluent community as the basis for the analysis. Even if the study's conclusions are, in fact, accurate indicators of the losses to be incurred without eradication, one might well ask so what?—especially if the funds for the eradication program are paid for by *all* taxpayers and if the potential harm from the widespread use of toxic chemicals is inflicted on a much broader community, as well as on future generations.

Suppose the public was presented with the following proposition: Should general tax revenues pay for a pest eradication program that protects the property of a relatively few wealthy individuals who could choose alternative solutions, when that program will have unknown but possibly important long-term consequences on the general population? It is likely that most political bodies would respond with a resounding no.

Why is so little attention paid to the values underlying the issue of equity? Part of the answer is that equity does not lend itself to easy quantification. Yet many efficiency impacts are also impossible to quantify, as we have seen in the above discussions concerning the valuation of life and nonmarket impacts. Ultimately, I believe the neglect of equity (as well as the elusive long-term environmental efficiency issues) serve a specific political purpose: these omissions generally insure the conclusion that eradication is good policy.

Cost-benefit analyses, based on incomplete data and on certain critical values, must emphasize immediate benefits over long-term and less well understood costs. For this reason, they can be used to justify policies desired by powerful groups having an interest in chemical forms of pest control. Unfortunately, the public often accepts the figures resulting from this approach, so such analysis tends to obscure the real issues and to rationalize bad policy.

Conclusion

I have concentrated on the many problems encountered in the application of cost-benefit analysis to pest management decisions. Some of these problems result from the practice of "bad economics" and are remediable. Some problems arise because we simply do not know enough about the basic pest management program, its eventual impact on the biological and physical environment, and its effects on the human social system. More study can improve our ability to anticipate the consequences of eradication programs, but the nature of these programs prevents sure, accurate predictions because they are generally carried out in locations where the pest is newly introduced and where much necessary biological information is not available. Under these circumstances I do not believe it will ever be possible to conduct adequate evaluations of eradication programs.

Finally, I have pointed to problems inherent in cost-benefit analysis that are essentially political and ethical. What weight do we give future generations in our planning process? How do we value human life? What is the value of the existence of an endangered species? What weight do we give efficiency when it conflicts with the goal of equity? These questions cannot be settled in an academic context; they are resolvable only within a political one.

Cost-benefit analysis was hailed as a tool that would force more rational government policy. In some instances, such as water policy (for which this tool was first developed), the tool provides useful information that could rationalize policy if the politicians and bureaucrats were forced to follow its conclusions. But the use of cost-benefit analysis in the evaluation of eradication programs is probably an "irrationalizing" rather than a rationalizing force, given our present understanding of the problems associated with such programs. Built into this tool is a powerful set of biases that favor the immediate and tangible results over the future and uncertain events. This bias is reinforced by lack of essential information.

Given that eradication depends on the use of toxic chemicals with widespread and long-term impacts, there is probably no such thing as a safe program that poses no significant risks. The alternative to eradication is management of the pest population. This alternative is generally pictured as less desirable because it accepts the notion that we can live with the pest and keep its population low enough to minimize adverse impacts. Pest management, however, allows time for study to determine how detrimental the pest really is and how it can be managed without imposing the costs and risks associated with widespread application of toxic chemicals. I would suggest that our ability to manage in this way is more developed than our ability to eradicate. Furthermore, if we are going to make mistakes, I would prefer those that result in a larger than desirable pest population rather than those that would place present and future generations at high risk.

My own conclusion is that cost-benefit analysis is more dangerous than helpful in the overall evaluation process where eradication is involved. By contrast, I think that economic evaluation of ongoing pest management programs is potentially very useful, mainly in helping to sort out some of the consequences of various alternatives.

But even in this more useful role of assisting in evaluating alternative management strategies, cost-benefit analysis alone cannot, in my opinion, produce definitive criteria for public policy. The problems of value cannot be overlooked or overcome, no matter how good our information base may be. There is no alternative to a fully informed political debate of the issues.

By *political debate* I do not mean what goes on in the legislature. Political debate is a broader term that implies widespread citizen concern and public participation. Such debate requires the enfranchisement of many interests not usually present in the policy-making process. In recent years, we have seen environmental groups become more visible and powerful in shaping the issues. This process must be expanded to encourage still broader participation. More information is the one critical ingredient for such participation to be meaningful. Information is needed regarding the unseen but potentially disasterous effects of expanding the use of toxic chemicals in the environment; it is needed regarding the effectiveness of past eradication efforts; it is needed regarding the distribution of winners and losers; and it is needed regarding the widespread abuse of cost-benefit analysis.

4
Institutional Aspects of
Eradication Projects

Olaf Leifson

This chapter will cover institutional aspects of pest eradication projects in California. My assignments in the state's Department of Food and Agriculture have included responsibility for environmental monitoring of pesticide use by emergency pest-eradication projects. They have provided relatively close contact with actual projects without direct administrative authority or responsibility for them. For this reason, the view presented here will be relatively detached and is also a personal one. Much of my discussion will be closely linked with those on public policy and legal considerations, and the separation is often a question of personal taste. This will not be a discussion on entomology, nor will it be pest-specific. My intent is to provide a feeling for the *institution*—the traditions, agencies, rules, laws, players—in other words, the whole social, political, and intellectual paradigm connected with eradication projects in California. I have organized my subject into the following segments: (1) a simplified interactional network to demonstrate linkages, (2) the traditions, rules, and agencies involved, and (3) the pest-prevention system.

Interactional Network

Perhaps the most effective way to begin to understand the institutional aspects is with a simplified diagram to illustrate the interactional linkages (fig. 4.1). My use of this diagram is motivated by a format used to illustrate pest control in its cultural and environmental contexts (Perkins and Pimentel 1980).

As I present this interactional diagram, several clarifying comments are relevant. The goal of an eradication project is biological extinction of a given pest in California. In a practical sense this means achieving a low enough population of that pest species in California to escape detection for a predetermined period of time. Because of the possibility of reintroduction, eradication is not likely to be permanent; hence the participating

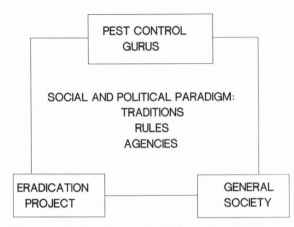

Figure 4.1 Functional interactional diagram for eradication programs.

agencies, individuals, and organizations have an ongoing, long-term commitment to establish institutional relationships and policy objectives. The interactional diagram represents a dynamic entity that continually changes according to perceived needs and resources, the abilities and aggressiveness of the individuals involved, and the constraints imposed by the project structure on the activities of project members. As a final observation, two factors are of great importance in this process: the way the organizational structure is established and the ways in which individuals function within and across structural lines in order to get things done.

The term *eradication project* here means a project that is generally chemically based. The information base, including methods, techniques, and fundamental knowledge, is provided by the pest control "gurus." The choice of this title is deliberate but is not intended to be flippant. Rather, it is to emphasize that the information base rests on a variety of sources, ranging from carefully controlled research plot results to anecdotal information of possibly dubious validity. The interpretation of this information for meaningful guidance in an eradication project requires many judgmental conclusions transcending pure science. Because of this sizable judgmental component, the organizational structure and the ways individuals function in various structures implicitly determine the context in which this process is carried out and whose judgment is used.

The term *general society* is used here to mean those segments of our urban or rural population and production agriculture most directly affected by the specific project. Providing the overall social and political paradigm within which the project operates are traditions, rules, agencies, and standards.

Traditions, Rules, Agencies

A short chronology of the California pest-exclusion concept includes the following significant dates and events:

April 5, 1880	Act to Promote State Viticultural Industries Passed
March 4, 1881	Board of State Horticultural Commissioners Established
1889	State Horticultural Quarantine Law Passed
August 20, 1912	U.S. Plant Quarantine Act Passed
1922	County Boll Weevil Inspection Stations Established
1972	Regional Exclusion Programs with Arizona Established
1982	Governor's Pest Response Task Force Established

The first organized pest-control activity, a disease control program for grape root rot, occurred with the 1880 passage by the California legislature of an act to "promote state viticultural industries." The State Board of Horticultural Commissioners was established the following year ("An act to define and enlarge the duties and powers of the board of state viticulture and to authorize the appointment of certain officers, and to protect the interests of horticulture and viticulture") and later evolved into the presently existing system of county agricultural commissioners. The essential role of the commissioners in any pest exclusion program will be described and needs to be emphasized.

The first state horticultural quarantine law was passed a hundred years ago, in 1889. This law underscores the continuing concern of the legislature for the protection of California agriculture for the last century. The law provided that a state horticultural commissioner was also to be deemed the state horticultural quarantine officer. In addition, the law required the county board of supervisors "upon petition of county freeholders to appoint a County Board of Horticultural Commissioners" and set forth provisions requiring transporters and receivers of plant material to report and hold shipments for inspection by the county, or the state horticultural commissioner.

An interesting facet of this legislation was that the state of California enacted a horticultural quarantine law that was also to be applied to shipments from foreign countries. Questions about the jurisdiction of state law over such shipments were resolved with the passage by the U.S. Congress of the Plant Quarantine Act of August 20, 1912.

By the early 1920s, because of the increased ease of automobile travel, pest introductions by tourists driving across state boundaries were noted. A quarantine station was established on old Highway 40 between Reno and Truckee to intercept alfalfa weevil. At that time there were few motels in the area, and motorists often camped in Truckee Meadows around Reno and the meadows to the east near Lovelock, Nevada. Alfalfa was grown in these areas and the "hitchhiking" alfalfa weevils caused enough concern in California to result in inspection stations on the California-Nevada border. In 1922, boll weevil inspection stations were established along the California-Arizona border for that serious pest.

From 1922 until the present, there has been a continuing development (subject to vicissitudes of budget) of these exclusion activities. Although they are not directly a part of a pest eradication project, they play an essential role in the total institutional structure within which any pest-eradication project functions. This chronology ends with the establishment by Governor Jerry Brown's administration of a Pest Response Task Force. Its function was to attempt to codify the lessons learned from the Mediterranean Fruit Fly eradication project in 1980–82. The traditions, then, go back about a hundred years.

The "players in the eradication game" are the agencies directly involved in some aspect of pest-eradication projects and are federal, state, regional, and county in scope. The key federal role is that of the U.S. Department of Agriculture, Animal, and Plant Health Inspection Service. Other federal agencies that may be involved are the Bureau of Land Management, Bureau of Reclamation, Corps of Engineers, and the USDA Agricultural Research Service. Within California, the lead is taken by the Department of Food and Agriculture and its divisions of Plant Industry and Animal Industry. The state attorney general has had an essential role in recent years. Because of pesticide regulatory constraints on these programs, the department's Division of Pest Management exercises a key oversight function. Other state regulatory agencies involved, because of potential environmental and human health impact, are the Department of Health Services, Department of Fish and Game, State and Regional Water Quality Control Boards, as well as the Air Resources Board and the Air Quality Management Districts. University research faculty, agricultural extension and agricultural experiment stations provide technical expertise. County agricultural departments under the direction of the county agricultural commissioners are responsible for the major portion of the eradication effort and the ongoing regulatory functions. An institutional aspect of an eradication program not discussed here is the very real difficulty of successfully combining county, state, and federal organizations into a coordinated project.

In addition to the governmental organizations, individual growers and agricultural trade associations play an important role in establishing pest eradication policy. These agricultural associations are influential in effecting agricultural policy. In a state as diverse as California, there are many such associations. A representative list includes: California Feed and Grain Association, California Rice Growers Association, Western Growers Association, California-Arizona Citrus League, California Tomato Growers Association, California Grape and Tree Fruit League, Agricultural Council of California, Calcot Limited, California Beet Growers Association, California Association of Nurserymen, San Joaquin Hay Growers Association, Cattlemen's Association, and Council of California Growers. This partial list illustrates the great diversity represented by the various associations.

The laws directly governing these projects are those in the Food and Agricultural Code of California. Other codes relating to health and safety, air quality, and water quality also constrain the actions of an eradication project. The California Food and Agricultural Code (FAC, Division 1, Part 1) provides for a state administration consisting of the Department of Food and Agriculture (DFA) with a director and several deputy directors who determine overall policy. There are explicit requirements: The DFA shall promote and protect agricultural industry (FAC, Section 401). It shall prevent the introduction and spread of injurious insect or animal pests and noxious weeds (FAC, Section 403). There is a provision that "with prior approval of the Department of Fish and Game and the Department of Health Services, the Department of Agriculture may reproduce and distribute biological control organisms that are not detrimental to the public health and safety and which are known to be useful in reducing or preventing plant or animal damage due to pests or diseases" (FAC, Section 405).

Other provisions of the Food and Agricultural Code directly affecting pest eradication projects are those dealing with pesticides and their use (FAC, Division 6). These provide a detailed statement of legislative intent together with the requirement that DFA formulates a pesticide regulatory program conforming to that intent to serve as a functional equivalent of the Environmental Impact Report-Negative Report required by the California Environmental Quality Act. Prior to adoption of the program, functional equivalency must be certified by the resources agency director. The current program was so certified by Huey Johnson, agency director from June 1977 to January 1983.

Regulations (Title 3, California Administrative Code) provide the details for implementation of the laws. The regulations are state agency rules filed with the secretary of state pursuant to the California Administrative Procedures Act. There are thus two fundamental codes governing an eradication project. One is the structure of basic law, and the other is a set of regulations administratively adopted by departments pursuant to exercising the authority assigned in the law.

The laws and regulations governing pesticide usage govern the way a pest eradication project can use chemical control technology. These laws provide (FAC Section 11501) for proper, safe, effective, and efficient use of pesticides while protecting the environment. They provide for safe working conditions for agricultural workers. They also generally encourage the development and implementation of pest management systems stressing biological

and cultural pest-control techniques using selected pesticides when necessary to achieve acceptable levels of control with the least possible harm to
nontarget organisms and the environment. The State Board of Food and
Agriculture reviews DFA programs and inquires into the needs and func-
tioning of the department. The state board consists of members chosen
from the agricultural and university communities (FAC, Division 1, Part 2).

The Code also provides for local administration (FAC, Division 2) consisting of a county department of agriculture administered by the county
agricultural commissioner. The commissioner is the county official with
primary responsibility in all the agricultural regulatory programs including
pest eradication projects. Our present California county commissioner system developed from that of the horticultural commissioners established
over a hundred years ago.

County agricultural commissioners and their deputies are chosen from
department certification lists based on satisfactory completion of a series of
examinations. They serve subject to review by the state director of food and
agriculture. Commissioners are individually appointed by the county board
of supervisors. Commissioners and their staffs are paid primarily by the
county with some state assistance. The county commissioner system insures
local flexibility while at the same time review by the director and interaction
with the DFA technical staff provides coordination and technical and administrative resources not normally available in every county. The complexity of
the county-state interaction is nowhere more fascinating to observe than
with the agricultural commissioners. The commissioners provide a direct
and influential input from local growers and agricultural commodity
groups to policy-making levels within DFA. Since commissioners often
have a longer job tenure than governors and food and agriculture directors,
abrupt policy changes from one administration to the next tend to be
mitigated.

In addition to state and local administration, the Code also provides for
plant quarantine and pest control activities (FAC, Division 4). These include:

1. Certification of plant shipments
2. Reporting of newly discovered pests
3. System of quarantine and inspection stations
4. Required abatement actions
5. Eradication projects conducted by county/state
6. Mandatory host-free periods for specific pests
7. Specific pest-control activities
8. Pest-control districts

Pest-Prevention System

Pest-control activities can be conceptually divided in four related parts that
together constitute our present control system. These parts are: (1) exclu-

sion via border station (DFA and federal, at ports of entry) and terminal point inspections, (2) detection, (3) eradication, and (4) control. An eradication project will involve all of these activities, directly or indirectly (fig. 4.2).

In order to see how an eradication project might come about, we begin with the statewide trapping detection system. This system attempts to find those pests that appear on a "hit list"—essentially a rating of pests according to their potential economic impact if established in California. Those pests designated an A-rated pest *shall* be eradicated by the County/DFA. An A-pest is an "organism of known economic importance subject to state (or commissioner when acting as a state agent) enforced action involving eradication, quarantine regulation, containment, rejection, or other holding action." A B-pest is an organism of known economic importance subject to eradication, containment, or control at the discretion of the individual county agricultural commissioner. A C-pest is an organism subject to no state enforced action outside of nurseries, except to retard spread. County action on a C-pest is at the discretion of the commissioner. Sometimes a

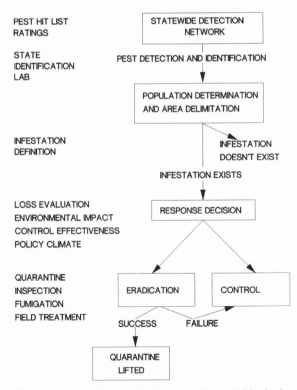

Figure 4.2. Flow diagram showing the various activities in the pest eradication system in California.

new organism may be found with either incomplete identification or inadequate information that *might* require an A-rating. Such situations are given a Q-rating pending completion of the evaluation. An example of a Q-rating occurred with the finding of carob moth in several Riverside County date groves in 1983. Pest rating evaluations are performed by a group of commissioners, affected agricultural industry members, and DFA specialists.

Given a pest hit list together with the statewide detection network, the system operates as follows: Once a pest is trapped and a preliminary identification made, it is sent to the DFA Identification Lab for confirmation. The next step is to deduce the total population and its extent. If the criteria for declaring the existence of an infestation of the pest are met, then the next steps are to: (a) perform a loss evaluation, (b) compare costs and effectiveness of eradication options, and (c) evaluate the environmental impact of each option. The conclusion of this analysis is a recommended response decision presented to the director for the final policy decision.

A response decision can be eradication, control, or no action. In the first case, an eradication project is established that functions within the traditions, institutions, laws, regulations, and agencies sketched in this discussion. The eradication project consists of several functionally distinct elements. These include an interior quarantine within the state boundaries, a commodity treatment system, and a field treatment program. This last aspect of eradication projects perhaps receives the greatest publicity, particularly if an urban area is to be treated.

The eradication project, consisting of the activities above, may at some stage be declared a success based on trapping results, may be judged too expensive to continue, or be judged to have failed. The last two of these possible conclusions are not the usual ones for a DFA eradication project, but they could occur and would signal a transition to a control mode. If the eradication is declared successful, then quarantine and postharvest treatments are terminated at the appropriate time.

Declaration of success for an eradication project is primarily based on trapping results and pest biology. But the influence of other governmental jurisdictions cannot be ignored. Ultimately, success or failure of an eradication project is determined by whether a state or foreign country will allow transport of unfumigated produce from the infested area. Such decisions may be influenced by more than numerical pest counts in detection traps.

In sketching the institutional structures within which an eradication project in California functions, it is important to remember that each project will be different. The particular circumstances of a given project are rarely repeated because the pest and the threat it poses, as well as policy constraints, vary considerably. The most effective pest eradication program does not necessarily mean following what worked the previous time. The most effective program has the greatest possible input from the gurus and effective linkages between the university, USDA, agribusiness, the general public, and the Department of Food and Agriculture.

Literature Cited

Perkins, J. H., and D. Pimentel. 1980. Society and pest control. In *Pest control: Cultural and environmental aspects,* ed. D. Pimentel and J. H. Perkins, 1–21. AAAS Selected Symmposium 43. Boulder, Colo.: Westview.

5

Legal Implications of
Eradication Programs

Charles W. Getz, IV

Not so long ago, an insect was something one simply squashed—eliminated, without giving the matter a second thought. At least, that was my understanding of pest control and entomology. Since then, as a deputy attorney general for the state of California, I have been involved in several state eradication projects and become an instant "expert" on the sciences of entomology, biology, population control, and chemistry, with a smattering of toxicology and pathology thrown in for good measure.

To entomologists and scientists examining the problems of attempted eradication of an exotic pest in an environment, there would seem to be no legal implications to their actions. But the history of eradication projects in California has shown otherwise. Indeed, since the famous Medfly infestation of 1980–82 the courts have increasingly been used as arbiters of controversies between those attempting to eradicate an exotic pest and those resisting eradication efforts.

Thus, challenges to eradication projects are a recent phenomenon, and that phenomenon is still in the process of being defined. There is (at least at this point) no body of law unique to eradication projects, or perhaps even to entomology. The variety of attacks on eradication projects has ranged widely, however, implementing traditional as well as more recent laws. An examination of these cases shows no pattern as yet.

One of the first legal complications in eradication efforts is that oftentimes layers of government are involved. For example, in California, the State Medfly Project was sued by Santa Clara County and several cities within the county in both state and federal courts, since the United State Department of Agriculture was also a part of that project. The state of California sued various southern states in the United States Supreme Court. California also supported private industry actions in a number of federal courts throughout the southern states. There were suits in California state courts between private individuals and the state. Thus, as one looks at an eradication program and its legal context, various laws and affected layers of government must first be examined.

On the federal level, the United States Department of Agriculture (USDA) has primary responsibility to ensure that exotic pests, especially those occurring in more than one state, are handled in a uniform, consistent manner. The major piece of legislation on such projects is the Federal Insecticide, Fungicide, and Rodenticide Act (FIFRA, 7 U.S.C. Sections 136 et seq.). This act was comprehensively amended in 1978, although the core provisions go back as far as 1912. Basically, the act regulates the use of chemicals and calls for registration of such materials with the Environmental Protection Agency (EPA).

Two other federal acts are also important—the Federal Plant Pest Act (7 U.S.C. Section 150aa–150jj), and the Federal Plant Quarantine Act (7 U.S.C. Section 167). These Acts authorize the USDA to declare quarantines. The Plant Pest Act specifically declares unlawful the importation or movement within interstate commerce of any plant pest without first obtaining a permit from the secretary of agriculture. As an emergency measure, the secretary can also quarantine or destroy any plant pest within the United States. The Federal Plant Quarantine Act (Plant Act) regulates the importation of nursery stock and prevents such plants from being moved in interstate commerce until a permit is obtained. Again, movements of plants in interstate commerce can be quarantined to prevent spread of a plant pest.

Under our federal form of government, states are bound to follow federal laws that "occupy the field"—that is, if the federal government enacts a law, where it is empowered to legislate, in the national interest, the states cannot then enact laws inconsistent with the federal law. Thus, where there is a federal quarantine, it would appear that states do not have the authority to alter that quarantine, *with the exception* of enlarging a federal quarantine area within their own borders. A state may not *reduce* the size of the quarantine area. In other words, a state may enact stricter laws than the federal government, but not less restrictive ones.

In the Medfly litigation, a major problem facing California was the attempt by several southern states, led by Texas and Florida, to prohibit state importation of *any* California produce from *anywhere* in California, notwithstanding the fact that the federal government had quarantined only the California counties actually infested with the Medfly. In two Supreme Court actions, California successfully overturned these attempted boycotts because of all the federal acts cited above.

On a state level, there are a variety of laws involved in an eradication effort; at first glance, only a few would appear to be applicable. In California the major body of law is contained in the State Food and Agricultural Code (a comprehensive set of laws governing state agriculture). Beyond the Code itself, three California Administrative Code sections (1 et seq.) contain regulations adopted by the department over the years to administer provisions of the Food and Agricultural Code. In the area of eradication projects, the director of food and agriculture has been granted broad authority by the legislature and indeed must, where feasible, eradicate any exotic plant or animal pest that poses a threat to California agriculture or the health and safety of its citizens.

These federal and state laws form the framework for eradication efforts and are continually being refined. From 1981 to 1983 I served on Governor Jerry Brown's Pest Response Task Force, created after the Medfly problem to critically examine pest-eradication efforts of the Department of Food and Agriculture (CDFA) and to recommend changes to its organization and governing law. I also served thereafter on the Pest Response Review Committee of the California Legislature, which continued the work started by Governor Brown's Task Force. The process of responding to needs within this legal framework in California is ongoing.

As mentioned earlier, the controversy over eradication efforts and resulting legal challenges is relatively new. Eradication projects themselves are not new—indeed, widespread eradication efforts using arsenic and other chemicals date back to the 1920s. Projects occurred as almost a matter of course throughout the 1950s and 1960s (Hagen, Allen, and Tassan 1981). In a more innocent era when environmental concerns were not so immediate and when scientific detecting measures and laboratory procedures were not so sophisticated, local citizens generally saw eradication projects as entirely beneficial. Very likely such projects would never be contested. For example, in 1960–61, Japanese beetle invaded the parklands adjacent to the State Capitol in Sacramento. This infestation was immediately fought with a combination of carbaryl and chlordane and was effectively eradicated. There was no public debate or apparent opposition to the project, even though the chemicals were freely sprayed in public parks. By contrast, in 1982, when Japanese beetle invaded a remote residential area in Sacramento County and a nearly identical chemical program was planned (using carbaryl and isofenphos (Oftanol), chlordane having been banned by the EPA), the local citizenry resisted strongly. There were court actions, physical threats, and public demonstrations against the project.

In the intervening twenty years there was a growing public awareness of the two-sided nature of chemicals—their benefits could be outweighed by their potential harm. There also was increased recognition that the environment had to be protected; this is reflected in the explosion of environmental legislation passed in the late 1960s and early 1970s. Armed with these new tools, many environmental groups and coalitions of ordinary citizens discovered they could resist government eradication projects and indeed, in some cases, felt it necessary to do so.

Legal Challenges to Eradication Efforts

The first truly organized judicial resistance to an eradication effort occurred in the Dutch elm project in 1977–78 in Palo Alto. Here was a community with many citizens well informed on environmental issues. The city did not want the state to treat an outbreak of Dutch elm disease with chemicals and filed an action stating that the California Environmental Quality Act (Public Resources Code Section 20000 et seq.) required a comprehensive environmental impact report (EIR) for any project having possible adverse environmental effects. The city claimed that the Dutch elm eradication project could not go forward unless such an EIR was prepared. While the case was pending the legislature amended the California Environmental Quality Act and created a new comprehensive pesticide registration program for considering the environmental effects of pesticides without preparing an EIR.

The Medfly controversy was the watershed for litigation and set the tone for litigation that followed. As mentioned, various cities and Santa Clara County sued the state of California and the secretary of agriculture in an attempt to stop aerial applications of malathion over Santa Clara Valley. The cases were based on several premises, including alleged violations of the California Environmental Quality Act. It was also claimed that local ordinances prohibited low flying aircraft from applying pesticides and that the governor's emergency decree did not negate compliance with state and local health laws that were allegedly violated by the project. None of the cases was successful.

In 1982, despite the preparation of an EIR for a gypsy moth eradication project in Santa Barbara County, the city of Santa Barbara and a coalition of environmental groups sued the state and Santa Barbara County in an attempt to block an eradication project. The project involved the use of *Bacillus thuringiensis* (*B.t.*), a microbial pathogen, and carbaryl applications to combat gypsy moth. The suits were based on the claim that the EIR was not "sufficient" nor properly prepared. One case was dismissed voluntarily; the other is still pending as of 1986, although dormant.

In 1983 the Japanese beetle eradication project was opposed on the theory that the use of isofenphos was a "public nuisance" and should not be allowed. Claims were made that state crews applying chemicals on private property were trespassing and should have landowner permission before going onto property. Also in 1983, citizens opposed a State Department of Boating and Waterways' project to eliminate water hyacinth. The project called for the use of integrated pest management, including limited use of the material 2,4–D as an interim control. Citizens in the eradication area claimed they had a right to a hearing before the county agricultural commissioner could issue a permit to apply 2,4–D. They further claimed that the California Environmental Quality Act required preparation of an EIR. None of the claims was successful.

A similar challenge to a 1984 gypsy moth eradication program in San Jose, which proposed the ground spraying of carbaryl, was unsuccessful in San Francisco Superior Court. However, in 1985 two suits opposing eradication projects were successful at the trial court level. The first was a suit brought by the county of Santa Cruz and a number of its citizens to stop a carbaryl ground spraying program (with some aerial spray in inaccessible areas) in the Felton area of Santa Cruz County. Inexplicably, an identical program to be carried out in Fremont was not opposed. In the six northern California counties, a proposal to treat an apple maggot fruitfly infestation with the insecticide imidan was challenged. A suit brought by citizens and an environmental group stopped that program. On appeal, the Santa Cruz decision was reversed, but the apple maggot case was affirmed.

Both of these suits used similar tactics: (1) the massive initial filing of substantial declarations and documents that questioned the inherent safety of the utilized material (despite previous environmental studies of the material and its registration), and (2) court hearings that put great pressure on all parties and the judiciary to decide an extremely complicated case at the last minute. Both the Felton gypsy moth case and the Humboldt County apple maggot case involved legal complaints with as many as ten causes of action, thousands of pages of documents, and a multitude of different law theories—all to be digested by a court within a matter of days. The result was that no judge had the opportunity to review the material, nor could any continuance or delay be allowed because of the immediate biological timetable requirements. Since poisons are involved in such cases, it is understandable that the courts exercise caution. Both courts in the above cases chose a conservative approach and stopped the programs.

In response to these tactics, the legislature overwhelmingly passed AB 1525, a comprehensive set of procedures unique to California and applicable to pest eradication legal challenges. The purpose of AB 1525 is *not* to overturn any particular decision or to prevent citizens from legally challenging pest-eradication projects. The purpose is to prevent unfair use of the inherent delays in the court system. Under AB 1525, the director is required to issue *written* findings in a decision of action, and the opponents of that action are required to adhere to a fair schedule for filing their complaint and supporting documents and for allowing opposition to that complaint in a timely manner. Further, courts are given sufficient time to review the documents and are reminded that they are not to act as scientists but merely to review the record for any errors of law. It is hoped that this legislation will enable the courts to give a balanced and fair review to pest eradication challenges.

Notwithstanding AB 1525, in 1986 the Department of Food and Agriculture anticipated further eradication challenges. Another lawsuit challenging the ongoing apple maggot project was unsuccessful, in part because the court had time to thoroughly review all the papers submitted. The introduction of the Africanized bee, the invasion of the achrane mite among bee populations, and the continuing introduction of such exotic pests as gypsy moth and apple maggot fruitfly will guarantee continuing eradication efforts and, more than likely, continued eradication legal challenges.

In the legal framework of these challenges, are there common denominators? Most challenges come from local citizens who do not want their private property treated with or affected by chemicals. Some of the cases do involve environmental groups or local government bodies reflecting a broad citizenry, but the motivating force for the litigation has generally been an individual who objects to the use of chemicals over, near, or on a piece of private property.

The implications of such challenges can be disturbing. A person has no legal obligation to be concerned with another's property, but state government, on the other hand, has a legal obligation to ensure the welfare of every citizen. Often the legal challenge to an eradication effort is based on concerns that extend only to the boundaries of the challenger's individual property. Without demeaning the motivation behind such challenges, what about the problem created when an exotic pest spreads and the consequent use of chemicals is required elsewhere? Should one person's individual rights be allowed to cause a situation in which more pesticides are then needed in other locations? Ultimately the courts must weigh the effects of an action on the individual versus the effects of nonaction, or modified action, on the entire state. The question of individual concerns versus the concerns of an entire community as represented by the state or national government is not easily resolved.

A second problem facing the potential challenger to an eradication program is limitations of a law on actual judicial action. The environmental acts governing air pollution, water pollution, use of chemicals, and preparations of environmental documents are comprehensive, but ultimately the question is about the court's authority when an eradication program has been declared by government. Surprisingly, the court has limited authority. Eradication programs involve technical and scientific decisions and are considered to be quasi-legislative, meaning that the administrative experts responsible for formulating them have certain expertise and create programs meeting certain statutory requirements. The courts, therefore, will defer to the discretion of the administrative decision maker unless there is a "clear abuse of discretion" or the program is "arbitrary and capricious." Many challenges to eradication programs boil down to the challenger wishing to substitute his or her judgement for that of the administrative decision maker, something courts generally do not allow. Courts will examine the legal validity of the decision-making process and, if there is a legal impairment, will invalidate the decision. But courts traditionally will *not* direct that a decision be remade in any particular manner.

The limitation on judicial action, which is part of the separation of powers doctrine between the legislative, executive, and judicial branches of government, perhaps leads to the greatest misunderstandings and frustrations for anyone attempting to challenge an eradication program. It is not enough to argue (and no matter in how much good faith) that one's proposed solution is perhaps safer or better than the solution proposed by CDFA or USDA. Furthermore, in most cases the contestant's proposed solution has not been scientifically established as equally effective as the one proposed by CDFA or USDA. Therefore, using a frontal attack approach will not likely be successful.

Consequently, many contestants try to find some technical reason why the decision of a government department was improper or did not comply with some obscure regulation. This is an attempt to win by delay what cannot be won on merit. Because eradication efforts must compete with a biological time clock, a delay can be tantamount to success in challenging an eradication project. Therefore, opposition to eradication decisions is increasingly being based on theories that initially appear to have no connection with entomology, eradication, or agriculture. For example, Federal Aviation Administration regulations governing low altitude flights have been used as an argument against aerial applications of pesticides. Public nuisance or trespass theories and the use of public review and comment sections in pesticide registration regulations have been employed to contest the use of pesticides. I do not impune such behavior as improper, unfair, or unethical. These tactics, however, seem to indicate a climate of "them versus us" in eradication projects—a climate not evident in the past.

Where do we go from here? Despite fervent opposition from some elements of the public, many government agencies sponsoring eradication programs see those programs as absolutely necessary for the economic welfare of the agricultural industry and, more importantly, for the general health and welfare of the citizens of a state.

The theory behind eradication projects is that using small amounts of chemicals *now* to control potentially devastating pests prevents a greater need in the future for the individual use of chemicals. In addition to this approach, government agencies have become more receptive to integrated pest management techniques and less reliant on chemicals. In California, this receptivity is not due to litigation, but because many agencies recognize their responsibility to protect the environment.

In my opinion, the state of California has made tremendous strides in responding to legitimate public concerns. Because of the recommendations of the Governor's Task Force, the State Department of Food and Agriculture has created science advisory panels to advise the department on available options for fighting exotic pests. The department is open to review and inquiry by citizens and state agencies concerned with the protection of the environment (such as the Water Resources Control Board, the air pollution control districts, and the Department of Health Services). Decisions concerning eradication efforts have not been opened to public debate but have been open to consideration of public concerns about the use of chemicals.

The continuing role of eradication programs in California promises some interesting and exciting events. The state's notable success in already eradicating many exotic pests has clearly affected the state's agriculture, and as it is one of California's primary industries, benefits to agriculture extend to the entire state.

Literature Cited

Hagen, K. S., W. W. Allen, and R. L. Tassan. 1981. *Mediterranean fruit fly: The worst may be yet to come. Calif. Agric.* 35:5–7.

6

Public Policy Considerations: The Medfly Case History

Jerry Scribner

The California Medfly experience is a fascinating study of governmental response to crisis in a democracy. It is also a good model for examining how public policy considerations are weighed in the context of insect eradication programs. The 1980–1982 Medfly program is reviewed here with particular attention to the public policy constraints inherent in large-scale eradication programs. First, however, the assumptions underlying the American political process will be examined generally, followed by a brief look at the way governmental decision-making authority is delegated in a complex society.

Policy Constraints

The American economic system is founded on the assumption that competition between individual entrepreneurs pursuing naked self-interest and profit is the pathway to economic order and material abundance for all. Similarly, our political democracy assumes that competing political and ideological constituencies will do battle in the market place of ideas, and from this the common good will emerge. The Medfly experience, as an example of insect eradication in a democratic setting, illustrates both the weaknesses and the wisdom of these assumptions. The clash of farmers' interests (wanting spraying when it was biologically inappropriate) with urban residents (opposing spraying when it was clearly the only remaining hope) led to a resolution at each step that was politically sound and generally biologically sound.

Critical elements influencing public policy decisions include: (1) information (and its corollary, misinformation); (2) time; (3) access—to decision makers and to the ultimate decision maker, the public; (4) structural conflict; (5) personality; and (6) logistics. Each of these is briefly discussed here.

Information and misinformation. An eradication program requires eliminating an insect from an environment in which it is not normally present, and because it is not normally present, we typically know very little about its behavior under these particular conditions.

In the case of the Medfly in California, some scientists thought the pest would not become established in the Santa Clara Valley; other scientists were convinced that a typical winter would eliminate even a large summer infestation. Information vital to any eradication approach—such as insect behavior, mating habits, and population dynamics—was limited to research conducted in laboratories or in tropical settings. The applicability of these research findings in Santa Clara County was unknown and, for the most part, is still unknown.

The second important factor is knowledge about the short-term and long-term impacts of any particular eradication tool. In general, there is more that we *do not* know than what we do know. This is true of nearly every potential chemical, biological, or cultural technique that could be used. Moreover, the limited knowledge we do have is often subject to differing interpretations by scientists. Minor differences between scientists can be easily magnified into major policy disagreements when presented to the general public and political leaders by competing ideological constituencies.

Time. Information and access are problems common to all public policy decisions. But the time frame is frequently a different and significant problem in insect eradication. Many, but certainly not all, environmental policy decisions are made against a backdrop of relatively slow-moving forces. Soil erosion, air pollution, water consumption, engineering projects, and waste recycling are some typical examples.

The watchwords of eradication are "early detection" and "prompt treatment." Normal environmental impact studies take several years and are followed by exhaustive public debate over their adequacy in light of new information. This process is totally inconsistent with prompt eradication.

In addition, any environmental analysis has to focus on a moving target. Medfly populations can multiply manifold, as often as every thirty days, and are capable of rapid geographic spread (Hagen, Allen, and Tassan 1981). Temperature and locale also affect population dynamics; thus the environmental impact report for the Medfly infestation in Los Angeles, where there were four adults and one larval site, would necessarily be different from the one for Santa Clara Valley, where dozens of larval sites and hundreds of adults were captured. An environmental impact report for Santa Clara in 1980 would have been different from one done in January 1981, and also different from one done in July 1981.

Access. A third critical element is the method by which necessary information for sound public policy choices reaches decision makers. How does it reach individual citizens in a democracy? The answer, to a large extent, is through the media, which is, after all, an entertainment business. Newspapers written like textbooks do not sell well; radio and television newscasts must be colorful, quick, hardhitting, and information is reduced to short, catchy phrases. The assumption is that the public finds the thorough dissemination of accurate information boring and that people are "too busy" to be well informed on anything. Because political leaders and public policy makers are also busy, attention tends to focus on dramatic contentions rather than on the more mundane yet accurate analyses of issues.

Structural conflict. Many of the problems with the Medfly project and with decision making originate with our governmental system. In nearly every area, there are several levels of government that are democratically elected and ultimately responsible to the same people. There are federal, state, and county health departments; federal, state, and county education departments; federal, state, and county water agencies. Often these co-responsible agencies view problems and solutions differently. Overlaying this is the free press, which interprets both problems and potential solutions to the public.

For pest-management decisions, the best analogy that can be made for assessing risk-benefit in a democracy is the following: Imagine an insect pest is found in your neighborhood. Ask yourself, Could I get the residents of the fifteen or more contiguous homes to sit down together in one room and agree on the parameters of the problem, the entomology, and the best strategy to solve it? Could all of this be done cooperatively? With the Medfly problem, magnify the single neighborhood into eight counties and forty-four cities, each county and city with its own elected representatives, plus eighteen state assemblymen and eight state senators. In addition, all of Stanford University was in the Medfly spray zone, and the University of California campus at Berkeley was not far outside it. Within each level of bureaucracy (federal, state, and county), all the different agencies have to work together. The decision-making process can, indeed, be complex.

There are deep-seated structural conflicts between and within the primary pest-eradication agencies. With Medfly and other pest problems there has been a history of dissension within the federal bureaucracy between the Agricultural Research Service and the Animal and Plant Health Inspection Service (APHIS). There is also conflict between federal and state agencies. California is the only state using pest-exclusion border stations, and it is the biggest agricultural state in the nation. The state attitude toward USDA is sometimes condescending—not a sentiment that encourages cooperation. The CDFA and the California Department of Health worked very well together on the Medfly problem, but there is a long history of conflict between them over pesticide use and the sharing of data and other information. There is also conflict between state and county governments, between county and city governments, and between county departments over matters of jurisdiction, funding, and competency.

Personality. It is probably fair to say that anyone in the Bay Area speaking in a pronounced southern drawl would not be taken seriously on a local environmental issue. In fact, USDA had credibility problems during the Medfly project because its personnel were generally seen as outsiders, and this was a significant factor in the local resistance to its attempts to achieve successful cooperation between the various local agencies.

Logistics. Logistics—getting money, people, resources, and policy decisions to mesh—is certainly a pest-eradication area fraught with difficulties. A case in point is the Northern California Medfly campaign beginning with the release of sterile flies in July 1980.

The Northern California Medfly Project

The Northern California Medfly Project can be divided into three distinct phases. The first phase started in June 1980. The primary eradication technique was the release of one billion sterile flies. The second phase (from December 1980 to June 1981) was a massive fruit-stripping and ground-spraying strategy along with the release of three billion more sterile flies. The third phase (from July 1981 through summer of 1982) involved aerial spraying of malathion bait similar to that used in Texas in 1966 (Hagen, Allen, and Tassan 1981), plus the ground spraying of the insecticide diazinon around larval finds.

Sterile Fly Releases

In the summer of 1980 a total of about 300 million sterile flies were released from mid-July to late August. Another 700 million were released between September and December; however, these numbers were believed to be far too few to overwhelm the number of wild flies present in the area. Approximately 60 percent of the flies were released on the ground by roving release trucks; the remainder were released by aircraft. Scientists on the Technical Advisory Committee disagreed on whether aerial release or ground release was the most effective.

By August 1980, when it was clear that there were not enough quality sterile flies available to eradicate the infestation, the Technical Review Committee recommended limited ground spraying in the most heavily infested neighborhoods. Program managers attempted to spray 4,500 backyards with malathion bait. But state law required homeowner permission, and many homeowners were either unavailable or, if available, refused to give permission. As a result, only about half of the yards were sprayed, further limiting the effectiveness of the program.

By November 1980, the infestation had expanded northwestward into

Palo Alto and across the San Francisco Bay into Alameda County. A request for additional sterile fly shipments from the U.S. Department of Agriculture/Mexico Laboratory in Metapa was denied on the grounds that the flies were needed for the Mexican eradication program. Aerial spraying was proposed in December, but community opposition prevented its implementation. Program managers appeared before six local government city councils and county boards of supervisors seeking spraying approval and were turned down on each appearance.

On December 24, 1980, the governor declared a state of emergency and mobilized resources from six state agencies to carry out a massive ground-based program during the winter months. As a deputy director in the Department of Food and Agriculture, I was assigned to run the project and was given broad powers to assemble the necessary personnel and equipment.

The Ground Program

The ground program had four key elements: (1) stripping all Medfly host material on or near previous larval finds; (2) tightening the quarantine; (3) ground application of malathion bait; and (4) the release of sterile flies.

An abandoned local school became the headquarters for a personnel force that grew from 200 to 2,000 over the next several weeks. On the first Sunday in January, the Medfly Project placed full-page advertisements in local newspapers asking community residents to strip their fruit trees and to report any maggot finds. One week earlier, volunteer coordinators had been appointed to each city; phone banks and other community volunteer program planning had also been undertaken.

The program concentrated on a 50-square mile (12,950 ha) area formed by delimiting a one-square mile (259 ha) eradication zone around each previous larva find. State employees went door-to-door in this 50-square mile area leaving notices at each residence explaining the proposed stripping and ground-spraying program. There were 3,000 blocks and over 100,000 homes in the affected area. Included in the notices were phone numbers for residents to call, with separate numbers for fruit-stripping and ground-spraying. Employees in the ground-spray phone bank took calls and were careful to see that the right information went into each record folder for each neighborhood block.

Fruit stripping by state crews. On January 6, 1981, 400 members of the California Conservation Corps (CCC), led by Food and Agricultural Department inspectors, began going through targeted neighborhoods to strip fruit. This was done on a door-to-door basis, and residents, if at home, were first asked for their cooperation in the project. About 45,000 properties had host fruit and were stripped. The stripped fruit was mainly citrus, avocados, and persimmons, but fruit such as prickly pear cactus was also included.

The stripped fruit was put into 30-pound (13.5 kg) plastic bags, stacked by the roadside, and then collected by California Department of Transportation (Caltrans) crews. The Caltrans trucks took the fruit to one of seven dump sites and, under the supervision of agricultural inspectors, buried it beneath two feet (60 cm) of soil. During the five-week stripping program, approximately 45 tons (41,000 kg) of fruit were collected.

The CCC worked six days a week under all weather conditions. They were assisted by the California National Guard as well as by Caltrans. The CCC force grew to over 1,000 men and women during January; all were housed at the Santa Clara County Fairgrounds and fed by Department of Forestry fire crews. In order to accommodate the large number of crew members the daily schedule began at 4:30 A.M., when the showering and eating shifts began. The Santa Clara County Fairgrounds had few shower facilities, so the National Guard erected shower tents and dug a sump into the fairgrounds to collect runoff. The Medfly project repaired the fairgrounds at the end of the project.

In addition to the stripping and ground-spraying program, the Medfly project increased the quarantine surveillance. Signs were erected on the freeways, and inspections were increased at airports, fruit stands, flea markets, and other points where produce was sold or handled.

Ground spraying. The Cupertino Caltrans Yard was the center of the ground spraying program, which involved about 550 employees and 118 ground-spray rigs. These crews worked six days a week.

Ground-spray employees had blood tests seventy-two hours apart to establish a baseline cholinesterase level; they were given additional blood tests if there was any sign of illness. Ground-spray crews met at 7:30 A.M. for briefing and safety checking of all equipment. In addition, the Medfly project was inspected more than 1,100 times by Cal/OSHA (Occupational Safety and Health Administration) and the Pesticide Safety Unit of the Department of Food and Agriculture. Each night, a crew filled the gas tanks and readied the equipment for the spraying the next day. When daytime crews arrived the next morning, all equipment would be ready to roll.

As with the fruit striping, block record folders were kept for each of the 3,000 blocks and over 50,000 residences involved in the ground-spraying program. Each night a crew at Medfly headquarters logged in all spraying for that day and prepared assignment sheets for the 118 spray crews for the following day. Folders had to include any information received from residents who had reported medical problems or who had restricted access to property or voiced objections to spraying. From January to June 1981, 62,000 backyards were sprayed six times with malathion bait.

Backyards around larval finds also were sprayed with diazinon as a soil drench to kill larvae and emerging pupae. Use of diazinon began in late June 1981. Earlier the program had used fenthion in backyards, but this was abandoned in early February when it proved ineffective.

Since loquats are a prime host in Hawaii, this fruit was given particular attention in Santa Clara County. Sterile flies were released around the trees, and the trees were sprayed and stripped. Loquats did not appear to be a significant host in Santa Clara, however; during the spring no adult Medflies were caught near them, nor were any Medfly larvae found in this fruit.

Sterile fly release. The fourth element in the intensive winter/spring eradication program was the great increase in the release of sterile flies. From June to December 1980, about one billion sterile flies were released. In the next six months, this number was more than tripled.

The Medfly project initially received flies from the USDA/Mexico Lab in Metapa, Mexico, and the USDA/Costa Rican Lab, as well as the USDA /Hawaiian Lab. But these sources were inadequate for California's needs. In January 1981, the Medfly project sent a team to Hawaii with instructions to quickly build a new Medfly rearing facility in cooperation with the Hawaii Department of Agriculture. The job took eight weeks; by early April sterile Medflies began arriving at the project in California.

Medflies from Metapa were dyed red, and up to 100 million arrived per week. Medflies from the California facility in Hawaii were dyed blue, and arrivals averaged 65 to 85 million a week. Flies from Peru, which began arriving in December, were dyed yellow, and flies from the USDA/Hawaii Lab were dyed green. Different colors of dye were used so flies could be distinguished by origin and differentiated from wild flies caught in the traps.

The sterile fly pupae arrived in jugs or polyethylene bags in 6–60 million lots. Medfly project personnel met the shipments at the airports and rushed them back to Medfly headquarters where they were put into Kentucky Fried Chicken buckets or boxes, depending on whether they were destined for ground or aerial release. Each bucket held 3,500 to 5,000 sterile flies.

After three days incubation at 80°F (26.5°C), most of the flies had emerged from the pupae and were ready for release. They were loaded into Medfly release trucks and driven into the neighborhoods. Thirteen trucks made three trips a day, five days a week, during the spring. Flies were kept at the proper temperature in the vehicles so that they could be released in the best possible condition. The trucks had a three-person crew: a driver, a navigator to record release locations and the fly lots being released, and a person to release the flies.

Developments, June 1981–November 1982

During the intensive winter/spring program a total of forty-three wild Medflies were captured and identified as wild. All were females, and most were nonmated. One fly was captured in March, thirty-one between April 6 and May 6, two more in later May, and nine from June 1 to June 5. After June 5, despite intensely hot weather and ample host availability, no wild flies were captured for three weeks. Many people on the project believed that the winter/spring program had succeeded. However, on June 25, Medfly larvae began showing up in apricots in Mountain View, California. Whether the renewed infestation resulted from the release of unsterilized flies (supposedly sterile), or a failure to detect the presence of additional wild flies or a combination of both is still in dispute. Within a week, more than sixty new larval finds had been identified, concentrated in the Mountain View area. On July 1, five days after the first find, Harvey Ford, of USDA, Rich Rominger, the director of CDFA, and I appeared before the California legislature. We advised that aerial spraying appeared to be the only remaining option and that it should begin on July 13, 1981 if recommended by the Technical Advisory Committee and approved by Governor Brown.

Despite the overwhelming evidence of new infestation, there was widespread and strenuous objection to aerial spraying. The Technical Committee meeting on July 7 was attended by approximately 800 people, including many local public officials. Nearly all who spoke to the committee testified in opposition to aerial application of malathion.

The first witnesses were children and pregnant mothers. They expressed concerns about the health impacts of aerial spraying. Many others in the community shared these concerns. The Technical Committee meeting was adjourned at 4:30 P.M. and reconvened the following morning in Sacramento with Governor Brown. At noon, the governor announced that he would not order aerial spraying at that time and instead ordered an even larger ground eradication program.

Park rangers participated in notifying the public of eradication steps. Within forty-eight hours the number of government employees working on the Medfly project increased from 400 to 4,000. However, the Medfly infestation grew faster than the ground efforts to control it. The USDA threatened to quarantine the entire state of California unless aerial application was ordered. On Friday, July 10, Governor Brown ordered aerial application to begin the following Monday night, July 13, the original target date.

Immediately after the governor's order on Friday, the Medfly project printed 40,000 notices of aerial application at the CDFA printing shop in Sacramento and delivered them to Los Gatos during the night. The next day these notices were distributed door to door to residents in the first aerial spraying corridor. During the day, a special Medfly health advisory committee reviewed the notice and recommended that it be changed, concluding that the notice overemphasized the health risks of aerial spraying. At midnight on Saturday project personnel flew the revised notice to Sacramento where 110,000 copies were printed and returned to the Medfly headquarters in San Jose for distribution on Sunday. During the course of the Medfly Project over two million notices were delivered by hand, door to door, or by mail.

In addition, a telephone hotline was set up to answer call-ins from residents about the quarantine, the requirements for renewed stripping, and the aerial spray program. Over a quarter of a million telephone calls poured in during July and August. Radio advertisements also notified residents of the requirement to strip trees, the prohibition against moving fruit, and reported aerial spraying times. Two press conferences a day were held, one in the morning and one in the afternoon, to answer the public's questions and to inform the media on the status of the project.

Because of concerns about flight safety, a decision had been made to use helicopters over the heavily populated Bay Area. However, the aerial spraying program had great difficulty getting started. On Sunday, July 12, the U.S. Department of Defense reversed the base commander's decision allowing the spraying project to use nearby Moffett Naval Air Station. The aerial program was then switched to the next available airport, the San Jose Municipal Airport. On Monday morning, the San Jose City Council met in emergency session and canceled the use of any public airport under its jurisdiction for Medfly spraying. In addition, the county and five cities sued the project in state court seeking to enjoin the spraying. During the day while most project officials were in court, Caltrans and the aerial spraying unit constructed a temporary secret helicopter base in the Los Altos hills. At 5:00 P.M. the superior court turned down the request to halt the spraying and an appeal to the California Supreme Court for an injunction was denied at 7:30 P.M. The aerial spray program began a little after midnight on July 13, as planned.

Most of the spraying was conducted at night between midnight and 6:00 A.M. to reduce the number of residents and automobiles exposed to the bait spray. There was also less air traffic at night, and the cooler, stiller air made application easier. Only one helicopter was used the first night, and the thick, sticky, corn syrup bait jammed the pumps of the helicopter after only six passes. The project continued to be plagued by pump problems during the first week, but by the second week corridors were being sprayed on schedule.

To guide the helicopters along a straight flight line, Caltrans crews put engineering markings along various roads so that spotlight crews could move from mark to mark setting up a straight line of spotlight beams for helicopter target guides. Later in the program, electronic guidance equipment was substituted for the spotlight beams. The ground guidance crews were protected by both law enforcement officers and sand-filled dump trucks. A number of automobile accidents were caused by drunk drivers hitting the rear of the spotlight crew trucks and roadblock obstructions. In addition, there were numerous alleged instances of gunshots being fired at the helicopters. Helicopters were actually struck by gunfire on at least two occasions. There were no accidents during the spray program itself, but in October one helicopter returning after a spray run crashed and burned in a residential neighborhood, killing the pilot. There were no serious injuries on the ground.

The helicopters flew at an altitude of 300 feet, 200 feet apart (about 90 and 60 m, respectively). More helicopters were added to the program as the spray area expanded. At the peak, 1,300 square miles (3,400 km^2) a week were sprayed using two contractors with six helicopters each. A third contractor sprayed with fixed-wing aircraft in the rural areas of Santa Clara, San Benito, and Stanislaus Counties.

In addition to the aerial spraying, roadblocks on all major freeways were maintained from July 8 through October 1. The inspections were conducted around the clock and a total of 5.2 million cars were stopped during the three-month period. Most of the vehicles stopped were campers and recreational vehicles towing trailers or trucks. Over 100,000 confiscations of host fruit were made, and 1,000 citations issued. The roadblocks helped reinforce public awareness of the problem and the prohibition against moving fruit.

Despite the roadblocks and the initiation of aerial spraying on July 14, during August the Medfly spread to the counties of Stanislaus, San Benito, Santa Cruz, and probably Los Angeles. In addition, infestations were found in areas of Alameda and San Mateo Counties. A small area of Contra Costa County on the border of Alameda County was also included in the spray zone.

At its peak in late August, the spray zone covered over 1,300 square miles, which were sprayed weekly. In early October, 400 square miles were dropped from the spray program. Another 700 were dropped in November, leaving about 200 square miles, which were sprayed every three weeks over the winter. In April, when warm weather returned, the frequency was increased from every three weeks to every week, with some areas being dropped from the spray program. The last spray in the Bay Area was on June 29, in San Mateo. On June 25, 1982, a single male Medfly was trapped in Stockton, California. Aerial spraying of a 9.5 square mile (25 km^2) area around this fly started the next day and continued weekly for two life cycles. No further flies were found in this area.

The Medfly eradication plan calls for spraying for two life cycles, then intensive trapping for one more cycle. At the end of the third life cycle, eradication is declared, and the quarantine ended. However, intensive trapping (up to fifty traps per square mile [2.5 km²]) continues for a fourth life cycle. California plans to continue maintaining five traps per square mile throughout the state to guard against future infestations.

The 1980–82 California Medfly eradication project officially ended on September 21, 1982, when the last quarantined area in the city of Stockton was declared eradicated. Intensive trapping in the eradicated areas (primarily in Stockton and San Mateo Counties) continued until November 1, 1982.

The Eradication Project was an immense undertaking that included the release of over four billion sterile flies, the stripping of fruit in over 100,000 urban backyards, ground spraying with malathion, Diazinon, or both, and the aerial spraying of approximately 900,000 acres in forty-four cities and eight counties. Counting multiple applications, more than ten million acres were sprayed with 190,000 gallons of 91 percent malathion. At roadblocks more than five million automobiles were stopped and inspected; telephone hotlines handled over 250,000 calls; and more than 14,000 claims for automobile paint damage were filed against the state of California.

Conclusion

Understanding the technical and logistical problems of eradication efforts is critical because it is hard to communicate to the public the difficulties of reaching all infested areas. Certainly one appeal of aerial spraying is its simplicity. The cost, however, is high. The costs for just the helicopters and malathion bait were $20 million. The claims against the state for automobile damage totaled about $16 million, of which approximately $3.7 million were paid. Various other lawsuits sought $2 billion.

There was extensive environmental surveying of air, water, and the amount of malathion bait landing on monitoring cards. These surveys indicated that there was very limited individual exposure to malathion. There were fewer than sixty lawsuits filed claiming personal injuries resulting from the spraying. By 1988, none had been successful, and only a few remained in the courts. Many of the car claims were meritorious and were settled, as malathion does pit enamel paint.

Hearing the previous statement, the average person could well ask If malathion eats through car paint, what will it do to my skin? Another health-related question might be If malathion kills flies, then couldn't it also harm people? These aren't unreasonable questions, but explaining the complexity of chemical reactions is not easy, nor are scientific explanations always adequate in allaying fears. The public has no clear grasp of where malathion, as a toxic chemical, fits in with other chemicals that are frequently inhaled, eaten, or used in some manner in our daily lives. The degree of risk is extremely difficult to quantify to a lay public.

In public policy debates the key issue is information, and the debate is only as good as the input. With Medfly, the public was trying to evaluate the magnitude of the problem. Agriculture was saying, "it's terrible," but the public did not agree, especially when treatment meant widespread use of malathion. To add to the confusion, some agricultural experts implied that one could "could eat malathion for breakfast—no problem." On the other hand, antipesticide proponents were associating malathion with World War II nerve gas and using such terms as "highly toxic." Reading the charges and countercharges in the newspapers during December 1980 could indeed cause confusion and concern.

Like information, time is a key constraint in eradication programs. An insect problem is so dynamic that it changes overnight. Just two weeks delay in discovering the Medfly in Santa Clara was enough for another half generation to develop. An environmental impact report for July 1 could arguably be outdated by July 15.

Access to the public was a critical factor. How did the state communicate with the public? The state spent a million dollars on newspaper and radio ads and distributed more than two million leaflets door to door. Yet, no matter how much information is available, the public tends to read selectively. How much unsolicited mail goes unread? How carefully were the flyers from USDA or the California Department of Food and Agriculture read, if at all? Getting information to the public can be a difficult problem, and it can be expensive. The Medfly project had to buy air time to inform the public on eradication plans; thousands of flyers had to be printed and distributed; telephone hot lines had to be maintained. The cost of supplying information was substantial.

Access to decision makers was another critical problem, particularly in a confusing and fast-moving situation. Even at the level of neighborhood meetings, it was not easy to decide who should get to speak first or longest, who would record the minutes, whose opinion carried the most weight. At higher levels, politicians were under pressure to use their influence for one side or the other. The situation was a jumble of confusion.

The problems encountered in the Medfly project may well occur with future pest-eradication efforts. Imagine, for instance, that the gypsy moth invades San Francisco. What is known about how this pest would behave in this particular environment? What is known about the long-term impact on the environment and on human health from any possible eradication tools that might be used? If one were suddenly in charge of an eradication effort, what would be the most effective way to approach the city council, the mayor, the county board of supervisors, the state legislature, the farming community? What would be a reasonable amount of time for an initial study and for preparing an environmental impact report? What individuals would manage the project? How should the scientific advisors be chosen? What about public input into the decision making process?

Meanwhile, "Friends of the Gypsy Moth," "Citizens Against Pests," and "The Coalition for Responsible Eradication" would also be mobilizing. The news media would be alerted. Television, radio, and newspaper reporters would descend on the project director and begin asking urgent questions, demanding immediate answers. How prepared would we be to find the right solutions?

Literature Cited

Hagen, K. S., W. W. Allen, and R. L. Tassan. 1981. *Mediterranean fruit fly: The worst may be yet to come. Calif. Agric.* 35:5–7.

Eradication Case Histories

7

Japanese Beetle

Daniel J. Clair and Vicki L. Kramer

The Japanese beetle, *Popillia japonica* Newman, was first discovered in the United States in August 1916 at a nursery in southern New Jersey. This beetle was previously known to occur only on the large islands of the Japanese archipelago, where it was not considered a serious pest (fig. 7.1). In the eastern United States, however, conditions were different, and the beetle became a significant pest rather quickly (Fleming 1976). It is believed that the absence of important natural enemies and the extensive areas of exploitable habitat allowed this dramatic change in pest status.

Life History and Habits

The Japanese beetle is univoltine in the United States (fig. 7.2). In certain northern parts of its range, however, some beetles may require two years to complete development (Hawley 1944). In most areas, adults begin emerging from the soil in June, and peak emergence occurs in late July (Fleming 1972). Males emerge a few days before females (Hadley and Hawley 1934; Regniere, Rabb, and Stinner 1981b). The half-inch long adult beetle is metallic green with copper-brown elytra that do not fully cover the abdomen (Fleming 1972). A pair of white spots dorsally on the last abdominal segment and a row of five white spots of hair on each side of the abdomen distinguish the Japanese beetle from similar-looking scarab beetles. Japanese beetles rest on plants close to the ground before flying off in search of food plants.

Newly emerged females produce a sex pheromone that is very attractive to males (Ladd 1970). For example, nine virgin females placed in a trap caught 2,975 males and 23 females within one hour. Emerging females are sexually mature, carry an average of twenty eggs, and begin to oviposit immediately after mating, probably even before feeding (Regniere, Rabb, and Stinner 1979). Unmated females carry a lifetime attractiveness to males (Goonewardene et al. 1970). Both males and females may mate several times

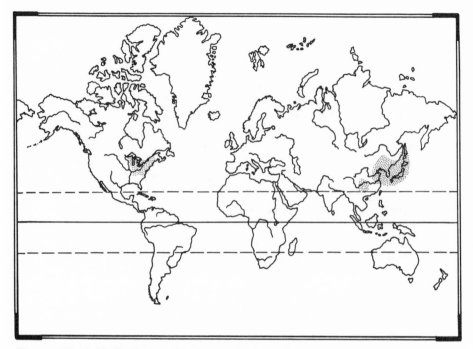

Figure 7.1 Worldwide distribution of the Japanese beetle, *Popillia japonica* (adapted from Commonwealth Institute of Entomology 1978).

and with several different partners (Fleming 1972). Mating usually occurs on plants in the early morning or early evening. Females continue to feed during mating, which may last several hours, but males do not. Females usually mate between successive oviposition periods.

Oviposition sites are selected on the basis of proximity to feeding sites, ground cover, and condition of the soil (ibid). Oviposition usually occurs near the area where the female has been feeding, preferably on grass-covered soil, with pastures being a common choice. Also, the soil must be moist to prevent egg desiccation and loose enough to allow digging. To oviposit, a female digs into the soil a few inches deep and deposits an egg in a depression made by the ovipositor. Further manipulation encloses the egg in a small cell of soil. Usually three or more eggs are laid in the same area. After ovipositing, the female may remain in the soil for up to four days, then return to a host plant to feed, mate, and reenter the soil to oviposit again. Females enter the soil a dozen or more times in their lifetime, laying up to sixty eggs. Adults live about thirty to forty-five days, females surviving a little longer than males.

After hatching, grubs begin to feed on nearby rootlets. They follow a

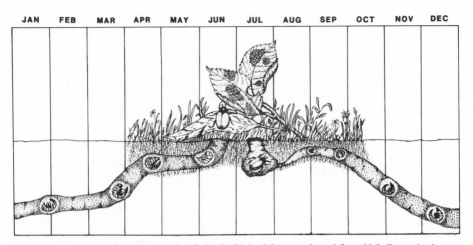

Figure 7.2 Life cycle of the Japanese beetle in the United States (adapted from U.S. Dept. Agric. 1973). Jan.–Feb., grub in winter cell. Mar.–Apr., grub comes up near surface to feed. May, grub forms cell and prepares to pupate. June, grub changes to pupa and then to adult, which emerges from ground. July, beetle lays eggs in ground, preferably in grass sod. Aug., eggs hatch; young feed on living roots of the plants. Sept.–Oct., grubs continue to feed and grow rapidly; injury to roots of plants is most common at this time. Nov.–Dec., grubs are mostly full grown and go to depths 4 to 8 inches below surface to pass winter in earthen cell.

rootlet until it has been eaten, then move horizontally through the soil until a new rootlet is found (ibid). Grubs always maintain a cell of space around them by digging with their mandibles and compacting the soil behind them as they move through the soil. Grubs also move vertically in the soil, being governed primarily by temperature and, to a lesser extent, moisture. In the summer most grubs are in the upper two inches of the soil. As the soil cools later in the year, grubs move downward until they become inactive at 10°C. Most overwinter as second or third instars at a depth between two and six inches. In the spring when the soil temperature warms to 10°C they begin moving back toward the surface.

Damage

The cryptic nature of the larvae often allows them to increase in numbers undetected until they become so numerous that their damage is obvious (USDA 1973). Unfortunately, at this point efforts at control may be too late to prevent economic loss. Not only are the roots of grasses in lawns and golf courses damaged, but the roots of vegetables and many ornamentals are also affected (Fleming 1972). In turf that is well maintained damage is usually not obvious until the density of larvae exceeds ten per square foot, whereas in poorly maintained turf the damage threshold is lower. Damage may be so severe that the dead lawn can be easily rolled back by hand because of root loss (Smith and Hadley 1926).

Odor is believed to be the primary cue for adult beetles in finding food sources (Major and Tietz 1962). Workers have shown that the odiferous compounds geroniol and eugenol, which are found in plants used by the species, are very attractive (Smith 1924; Langford et al. 1941; Fleming 1972). In addition, the presence of other beetles will also cause adults to land on a particular plant (van Leeuwen 1932). If a virgin female is present, many males will be attracted. Consequently, two plants of the same species in the same area may have very different numbers of beetles.

Hawley and Metzger (1940) found that plant feeding patterns change during the life of the adult beetle. Newly emerged beetles prefer low-growing plants for the first few days, then switch to fruit and shade trees for the next several weeks. When this latter food source becomes old and tough the beetles return to the low-growing plants.

In Japan there are 11 preferred plants of some 40 plants used by the beetle (Fleming 1972). In contrast, the United States has over 300 plant species that the Japanese beetle finds palatable (Hawley and Metzger 1940). Records on attack susceptibility are available for 435 plant species (Fleming 1972). Of these, 47 species are always attacked severely, 59 species receive moderate damage, and 67 species receive light damage. Some of the heavily damaged species include Japanese maple (*Acer palmatum* Thunb.), asparagus (*Asparagus officinalis* L.), American chestnut (*Castanea dentata* Borkh.), soybean (*Glycine max* [L.]), black walnut (*Juglans nigra* L.), crabapple (*Malus baccata* L.), apple (*M. sylvestris* Mill.), apricot (*Prunus armeniaca* L.), sweet cherry (*Prunus avium* L.), sour cherry (*Prunus cerasus* L.), peach (*Prunus persica* [L.]), nectarine (*Prunus persica nectarina*), poison ivy (*Rhus toxicodendron* L.), roses (*Rosa spp.*), American linden (*Tilia americana* L.), American and English elm (*Ulmus americana* L. and *Ulmus procera* Salisb.), sassafras (*Sassafras albidum* [Nutt.]), grapes (*Vitis spp.*) and corn (*Zea mays* L.). The beetles' preferred host is smartweed (*Polygonum* spp.), which grows in well-drained, light-textured soils, very suitable for oviposition (Regniere, Rabb, and Stinner 1983).

The Japanese beetle causes damage to orchard crops by feeding on the leaves and fruit. Many beetles may aggregate on the early ripening fruit of peaches or apples (Fleming 1972). For example, there is a record of 296 beetles being found on one apple. In corn, yield may be reduced when beetles feed on the silk before pollination occurs or burrow into the ends of ears and feed on developing kernels (Smith and Hadley 1926). Populations of beetles may reach extremely high numbers on crops. For example, Smith and Hadley (1926) recovered Japanese beetles from 10-year-old peach trees and found an average of 17,600 beetles per tree.

In 1955 damage by the Japanese beetle was estimated to cost $10 million per year in fifteen infested states (Fleming 1972), damaging approximately 16 percent of the country's major crops. During the thirty years prior to 1961 federal expenditures for control were close to $14 million, or about $500,000 per year. Nonfederal expenditures were almost $11 million for the same period. Industry reported that compliance with quarantine regulations cost them $500,000 per year.

Epidemiology and Ecology

In 1916 the Japanese beetle was known to exist in the United States only from an area of about one-half square mile in southern New Jersey. Fleming (1972) reports that from this initial infestation, the beetle spread eastward to the Atlantic Ocean by 1926 and south to the Delaware Bay by 1930 (fig. 7.3). In general, river valleys increased the beetle's rate of spread, whereas mountainous areas slowed its movements. Pennsylvania was first occupied in 1920, Delaware in 1926, New York in 1932, Maryland in 1933, Connecticut in 1937, Massachusetts in 1942, West Virginia in 1944, Rhode Island in 1946, New Hampshire and Vermont in 1951, and North Carolina and Ohio in 1952. The above dates represent times when the general area of infestation

Figure 7.3. Spread of the Japanese beetle in the United States from 1916 to 1952 (adapted from Fleming 1972). The numbers shown indicate the years that beetle infestations reached each state.

reached those states, although isolated infestations often occurred much earlier. Exact records were kept only until 1952, so figures since that time are estimates. By 1962, an estimated 100,000 square miles were infested, and probably about 150,000 square miles by 1970. Today the beetle is found as far west as Wisconsin, Illinois, and Missouri and southward into South Carolina and Georgia.

In 1920 Federal Quarantine 48 was initiated to control interstate movement of all agricultural products. States infested by the Japanese beetle also established state quarantines that regulated intrastate movement of agricultural products unless chemically treated. However, nonagricultural products were not under regulation, and therefore beetles could easily be spread by vehicles carrying these products. In an effort to prevent long-range spread, airplanes leaving infested areas were treated during the adult beetles' flight period.

The United States Department of Agriculture (USDA) placed monitoring traps near points where the beetle was most likely to be introduced, such as airports, truck depots, and freight yards (Fleming 1972). In 1930 the beetle was found at 61 isolated locations, the furthest being in Pennsylvania, about 285 miles from the nearest general infestation. By 1949 the beetle had been found at 570 locations, and by 1961 it was discovered as far away as California. Most of these isolated infestations were so small that they disappeared without being treated or were easily eradicated with insecticides applied to soil and foliage. Other infestations, however, became large and economically damaging. Japanese beetles have also been discovered several times on aircraft arriving in Europe from the United States. However, no infestations have yet been recognized in Europe (Fleming 1972).

The main factors affecting the spread of the beetle, besides food and oviposition sites, are topography, temperature, rainfall, and wind (Fox 1934; Hawley and Dobbins 1945; Regniere, Rabb, and Stinner 1981a). When extensive areas of flat farming land are present, the beetle spreads quickly. However, mountainous areas or forests slow the beetle's movements.

The beetle's eggs and larvae are susceptible to desiccation in the soil. For survival, there must be rainfall or irrigation throughout the year of no less than ten inches. In the summer the soil temperature must be between 17.5° and 27.5°C for development and survival. In winter, temperatures must exceed −9.4°C to prevent larval mortality. Snow cover can help to thermally insulate the soil and prevent mortality even when the air temperature is less than −9.4°C. Much of the eastern United States fulfills these requirements. In the central states, summers with less than ten inches of rain occur more frequently than in the eastern states, and much of the west and southwest is very dry. This will help prevent serious infestations in these parts of the country. The future spread of the beetle has been predicted by Fox (1939), Fleming (1963), and Hawley (1949), based primarily on temperature and rainfall requirements of the beetle. The northern limits of spread have been predicted to be in the higher elevations of New England and New York and in northern Michigan. The limiting factor most likely will be the low spring and summer temperatures, which will slow development so that oviposition may not occur before August. Eggs may hatch, but the larvae would most likely not survive. The area west of the Great Lakes to the Missouri River receives low snowfall and temperatures drop very low in the winter. The lack of an insulating snow cover combined with low temperatures apparently prohibits survival of overwintering grubs. West of the 100th meridian to the Rocky Mountains, summers are too dry for the beetle to survive, and where summer rainfall is sufficient, such as the higher elevations, temperatures are too cold. The Great Basin, between the Rocky and Sierra Nevada mountain ranges, is too dry, except along streams or irrigated areas, where the beetle could survive. In northern California, western Oregon and western Washington, summer temperatures are too cool for beetle development. Also, low summer rainfall along much of the Pacific coast would be prohibitive. However, extensive areas of irrigation in warmer areas could result in establishment, such as in Sacramento in 1961 and 1982 and San Diego in 1973. The mean summer temperature in southern Japan is 25°C. Temperatures above this in Florida and other southern states seem to have slowed the spread of the beetle southward (Fleming 1972), although earlier predictions were that the beetle would continue moving south.

Part of the problem in controlling the spread of the Japanese beetle is the beetle's frequent flights and its ability to fly long distances. In the morning, when the temperature reaches about 21°C and humidity is less than 60 percent, beetles begin to fly from the plants on which they have been resting to other plants (Fleming 1972). Most flights are for short distances, although flights of five miles have been recorded (Fleming 1958).

The annual spread of the beetle has been estimated to be between 10–15 miles (Smith and Hadley 1926) and between 2–10 miles per year (Fox 1932). In addition to the random individual flight, aggregates of millions of beetles have been observed leaving heavily infested areas in New Jersey. From 1923 to 1933 these mass flights occurred in late July or early August from rural New Jersey into downtown Philadelphia (Fleming 1972).

In the presence of chemotropic stimuli (primarily from plant compounds), flight is directed. Without these stimuli, however, flight direction is determined primarily by wind. Winds in southern New Jersey are usually westerly. During the first ten years of the beetle's infestation in the United States, movement occurred 42 miles to the east but only 28 miles westward (Fox 1927).

High population densities usually occur in areas where the adults have both food and oviposition sites (Regniere, Rabb, and Stinner 1983). However, because high grub populations are often found in areas distant from adult host plants, movement of the adults is important in the dynamics of this insect. Regniere, Rabb, and Stinner (1983) proposed a conceptual model of the beetle's movements based on four categories of sites, according to reproduction and survival potential in these sites:

1. intensive production sites, where adults can both feed and oviposit;
2. marginal production sites, where oviposition is sporadically suitable;
3. migration alleys, where conditions are usually unfavorable for oviposition and survival; and
4. feeding sites, where host plants occur isolated from suitable oviposition sites.

Control

There are many natural control agents of the Japanese beetle. Microbial pathogens provide the most effective control, whereas parasites and predators play a lesser role.

Milky disease, caused by the bacterium *Bacillus popillae* Dutky, was first noted in 1933 when a few abnormally white grubs of the Japanese beetle were found in central New Jersey. The bacterium grows and sporulates in the blood of living larvae, causing the normally clear hemolymph to appear milky (Fleming 1968). This disease has been shown to severely deplete the fat body cells of the larvae. Thus milky disease per se does not kill the larvae; rather, the larvae die because they lack the energy reserves necessary to proceed into the next metamorphic phase (Sharpe and Detroy 1979). Milky disease spores were first produced and distributed as part of a U.S. government program and later by private industry. Two U.S. companies, Fairfax Biological Laboratory and Renter Laboratories, are producing the bacterium under the trade names Doom, Japidemic, and Milky Spore.

Milky disease is known to persist for many years in the soil. The bacterium was applied extensively in Connecticut in the late 1940s. The beetle population declined and for some twenty-five years was an inconsequential pest in Connecticut. In 1973–74, however, heavy infestations of Japanese beetle caused severe damage throughout the state. A 1974 study showed that milky disease was still present in nine of sixteen of the original inoculation sites but that it was not producing the incidence of disease and control once observed. This suggests that the bacterium had weakened over time so that it was less invasive or that the larvae had developed some resistance to the disease (Dunbar and Beard 1975). Studies in New Jersey and Delaware in 1974 showed that milky disease was still present and providing effective control of the beetle (Hutton and Burbutis 1974).

Bacillus thuringiensis (Berliner) var. *galleriae* (*B.t.*) is another bacterial pathogen found to be effective against Japanese beetle grubs. This is the first evidence that *B.t.* is effective against a Coleopteran species. It is not a natural pathogen, as it has never been identified as a cause of death in the field. Practical application of *B.t.* to control larvae in the soil would require protection for the endotoxin crystal. Data indicate that application would be most effective in spring and again in late summer when the larvae are actively feeding (Sharpe 1976).

There are a number of predators on the Japanese beetle. Insectiverous birds and domestic fowl prey on adults; moles, shrews, and skunks prey on grubs. The activity of predators is, however, too restricted and too sporadic to have much effect on the population within a region (Fleming 1976).

Many parasites have been imported from the Orient but only two parasitic wasps, *Tiphia popilliavora* Rohwer and *T. vernalis* Rohwer, have become established in this country. These parasites are effective only against dense grub populations. Apparently it is difficult for the wasps to find the larvae in the soil when populations are low (ibid).

The nematode *Neoaplectana glaseri* Steiner was found attacking grubs in New Jersey in 1929. With optimum soil temperature, moisture, and dense grub populations, the nematode can parasitize up to 81 percent of a larval population. But, in general, *N. glaseri* is not effective throughout an area infested by the beetle (ibid). A mermithid nematode, *Psammomeris* sp., was recently discovered in the northeastern United States. The nematode has been reported to kill between 4.5 percent and 60 percent of the grubs (Klein et al. 1976).

Various cultural control practices can also be utilized to minimize the impact of Japanese beetle populations. Resistant or tolerant plants can be cultivated, particularly varieties of grapes, apples, peaches, and soybeans. In addition, early or late planting of crops to avoid the major feeding period of the adult and systematic hand removal of the gregarious adults can reduce economic damage. Crop rotation such as corn-soybean and tilling the soil by plowing and discing also reduce beetle populations (Fleming 1976).

Other techniques, such as mechanical and sex-pheromone traps, reduce adult numbers and thereby curtail the laying of eggs. The release of adequate numbers of sterile male beetles is also possible and could stabilize or reduce beetle populations in isolated areas.

Chemical control has been used over the years. The first major synthetic materials employed were chlordane and dieldrin for grub control. Because of the adverse environmental effects of these chemicals on nontarget organisms they are no longer registered for beetle control (Tashiro, Bourke, and Gibbs 1981). In addition, however, they were no longer effective in many areas, as the beetles had developed resistance to them. Since the early 1970s the organophosphates have been the most commonly used insecticides; fensulfothion and isofenphos have been shown to give excellent control of grubs (ibid; Lawrence 1981).

The economic damage caused by this beetle in the mid-Atlantic states has declined greatly over the last several decades, and at present severe plant damage is limited to a few isolated areas. It is believed that the complex of indigenous and introduced natural enemies discussed above was the principal reason for this general decline in severity (Fleming 1976).

History of Eradication Efforts

Eradication efforts were begun against the Japanese beetle soon after its introduction into the United States in New Jersey (ibid). A survey in 1917 showed that the original infestation covered approximately 2.5 square miles. Even though this was a small area, eradication of the beetle was difficult for several reasons. First, the beetle is a strong flier and can spread quickly. Second, its wide host range allowed it to find food easily rather than be killed by the insecticides and fungicides used. Third, arsenicals did not work well, and the beetle was repelled. Fourth, the larvae live in soil and are hard to find and difficult to reach with insecticides.

In the summer of 1918 a one-half to one mile wide barrier zone was formed around the infestation. Noneconomic host plants in this zone were cut and burned or coated with a repellent of lead arsenate and lime. Inside the infested area boys were paid eighty cents per quart to collect adult beetles by hand. Although 450 quarts were collected, this did not seem to make much difference. The beetles easily flew over the barrier zone, and the infested area increased to 6.7 square miles by the end of the summer in 1918.

In the fall of 1918 approximately 17 acres of pasture were treated with sodium cyanide. The efficacy of this treatment was variable. By the end of the summer of 1919 the infestation had spread to include 48 square miles, in spite of continued eradication efforts. That fall, 50 acres of pasture were treated with sodium cyanide.

In 1920 a final effort was made to eradicate the Japanese beetle. Host plants were dusted or killed with herbicides. Heavily infested orchards were sprayed with lead arsenate and beetles were collected by hand. By the end of that summer the infestation had doubled to 92 square miles in New Jersey and 11 square miles in Pennsylvania. The failure of the program resulted in its discontinuation.

Eradication efforts were in response to a public demand, although little information was available on the life history and habits of the beetle or on good control methods. For example, using a repellent on host plants did inhibit feeding on those plants but did little to stop the spread of the beetle and may have even increased it. The failure of the program did, however, produce an interest in studying the life history and habits of the beetle and ways to control it (ibid).

In June of 1932 and 1933 several Japanese beetle adults were collected in a backyard in St. Louis, Missouri. In February 1934 the Missouri Department of Agriculture notified the Bureau of Plant Quarantine of these collections. Investigation of these collections was begun immediately, and plans were made to trap and survey the area where the beetles were found and other likely surrounding sites (Stockwell 1935).

In 1934 a total of 2,600 traps were sent to St. Louis, along with an experienced trap supervisor and two trap scouts. That year a survey was made of all nurseries and greenhouses within a ten mile radius of St. Louis. The only infestation found was confined to a 117-block area within the city. Irrigated lawns and gardens in this residential area provided suitable habitat for the beetle.

Plans were made to apply lead arsenate to the entire infested area plus a safety zone. Labor was supplied by the Missouri Relief and Reconstruction Commission, while funds and equipment were obtained from the Missouri Department of Agriculture, the Federal Bureau of Entomology and Plant Quarantine, and the city of St. Louis (Dawson 1936). Experienced personnel were also sent from Harrisburg, Pennsylvania, to assist and operate ten federally-owned high pressure sprayers.

The Missouri Department of Agriculture notified all property owners and tenants about the program and asked for their cooperation. Treatments began September 26 and ended on November 8. A total of 440 acres was treated at a rate of 1,000 lbs of lead arsenate per acre. Even with this extremely high application rate, there was very little opposition from homeowners and tenants, undoubtably because of the general lack of knowledge regarding pesticides at that time (Dawson 1936, 1937, 1938).

Great effort was put into eradicating the Japanese beetle in St. Louis because it represented a threat to the agriculture of the midwest. In addition to control by lead arsenate treatments, a state quarantine was put into effect to stop movement of plants and plant products out of the quarantined area.

Ten thousand new and more efficient traps were used in 1935. Numbers of beetles trapped in heavily infested areas were treated with a total of 48 tons of lead arsenate, and the area under quarantine was extended to include newly infested areas. The conclusion that year was that the results were encouraging and invasion of the beetle to the midwest would be delayed for a long time.

In 1936, 4,000 of the older type of traps were added to the 10,000 newer traps. The beetle population showed a great reduction with numbers of beetles trapped dropping from 1,351 and 1,232 in 1934 and 1935, respectively, to only 88 in 1936. A total of 31.6 acres were treated in 1936 with 15.8 tons of lead arsenate. Another 13.5 tons of lead arsenate were applied to areas where construction projects had disturbed the soil treatments of the previous year. Vegetable and flower gardens were treated with double strength carbon disulfide emulsion, since lead arsenate would be ineffective in soil that was disturbed each year. The conclusion was that eradication seemed possible if results continued to be as good as in the previous three years.

In 1937, the trapping program used 14,750 traps and caught only one beetle. The catch was near a freight yard in an industrial part of the city. It was therefore thought that this find was probably unrelated to the city's infestation. The area around this find and a nearby garden were treated with lead arsenate and carbon disulfide emulsion, respectively. Hand scouting of previously infested areas, as a check to the traps, found no beetles. The trapping program was continued in 1938 to insure that the Japanese beetle had indeed been eradicated. No information was provided concerning the adverse environmental impact of the high chemical dosages on the treated areas.

Langford et al. (1941) reported on a cooperative program that began among the University of Maryland, the Bureau of Entomology and Plant Quarantine, and local county governmental agencies in Maryland to retard the spread of the Japanese beetle in that state. This program points out some of the differences in strategy when the goal is containment rather than eradication. No single method was effective by itself in controlling the beetle. Chemical control was too expensive for almost all homeowners and farmers; consequently, biological control was stressed. This was supplemented by education and demonstrations of cultural and such other control methods as crop substitutes, delayed planting, and plowing, discing and harrowing the soil to kill grubs. Biological control was confined to the use of milky disease and two parasitoids, *Tiphia vernalis* and *T. popilliavora*. The use of large numbers of traps was thought to reduce crop damage in most cases. For example, in one year 127,122 traps caught 275 tons of beetles.

Fleming and Hawley (1950) reported on a test program designed to evaluate the efficacy of DDT in controlling or eradicating the Japanese beetle in isolated infestations. The test was done in Blowing Rock, North Carolina, a resort town in the northwest of the state at an elevation of 3,600 feet, consisting mostly of hilly areas wooded with evergreens. There were several hotels, an 18–hole golf course, many small farms and residences with well-kept lawns. Rainfall and fog were common in the summer, but the temperature was low and winter temperatures were often below freezing. The beetle was introduced sometime prior to 1943, the year that more than 17,000 beetles were caught in just a few traps. That fall approximately 25 acres were treated with lead arsenate at 500 lbs per acre. In the summer of 1944 this treatment was shown to have had little effect on controlling the beetle.

In 1945 the entire infested area of 251.8 acres was treated with DDT, including the golf course, pastures, lawns, gardens, and flower beds. The DDT was applied at 25 lbs per acre active ingredient as a 10 percent dust (250 lbs per acre). The program was evaluated by soil sampling and trap captures, the latter being the more accurate measure. In the untreated area surrounding the infestation, the number of beetles caught increased each year from 357 in 1945 to 16,798 in 1948. In the treated area the number of beetles captured decreased from 138,945 in 1945 to 15,908 in 1948, a reduction of 89 percent. It was expected that this number would continue to decrease due to the residual effect of DDT.

Observations were made on the effect of the treatment on nontarget organisms. Since these were merely casual observations, the conclusions are not necessarily valid in all cases, especially concerning long-range effects. Carabid beetles were essentially eradicated, whereas most other insects were thought to reestablish. Flies were reduced in number in 1945 and 1946. Shortly after treatment, many nontarget animals were killed. For instance, insects above ground were killed, and due to this many toads and birds were killed or injured. There was no evidence of damage to fish, frogs, or other aquatic life. The conclusion of this project was that treatments of soil and foliage with DDT inhibited normal increases in population levels. The treatments would, however, have to be reapplied to disturbed soil, and the treatment zone would have to be expanded to pick up developing infestations.

In 1953 an infestation covering 10,000 acres of farmland was discovered in Illinois along the Indiana border (Luckmann and Decker 1960). The situation was considered dangerous due to the extent of the infestation, the value of agriculture in the area, and the fact that corn and soybeans were the main crops, both of which are attacked by the beetle. The program was operated by the Plant Pest Control Division of the USDA and the Illinois Department of Agriculture.

Treatments began in April 1954 with aerial application of dieldrin spray and granules at 3 lbs active ingredient per acre. Due to the hazards of the spray only granules were used after the first year. Roadsides within infested areas were sprayed with one lb. of DDT active ingredient per acre. All areas were treated only once. From 1954 to 1957 a total of 9,649 acres in Illinois and 3,460 acres in Indiana were treated. In 1958 a one-half mile wide strip on three sides of the infestation, excluding the Illinois-Indiana border area, was treated. This band was three to five miles from the outer edge of the infestation and covered 8,195 acres; thus, a total of 17,844 acres of Illinois farmland was treated during the five years of the program. Population density estimates were made by several means, including sweep nets, soil sampling, examination and dissection of plants and plant debris, and visual inspection. Traps were apparently not used.

Livestock was carefully monitored, since effects of dieldrin were unknown at the time. Milk samples were collected over time, and tissue samples examined from animals that died. Forage samples were taken to monitor residues. Animals exposed to dieldrin granules were not affected, but those exposed to drift from dieldrin sprays often became ill, and several sheep and many cats were killed.

Several common insects were reduced or eliminated for varying periods of time, while others increased in abundance, even though their natural enemies appeared unaffected. Still others, such as cicadellids and fulgorids increased due to the decrease in their dryinid parasitoids. The conclusion of this project was that, although dieldrin gave good control for five years in treated areas, untreated areas supported increasing populations, so that by 1958 the infestation had grown to 50,000 acres.

Gammon (1961) has reported on the first Japanese beetle infestation in California. An entomologist of the California Department of Agriculture first discovered the Japanese beetle in Capitol Park in Sacramento on June 7, 1961. Further searching turned up 14 more beetles, which confirmed its establishment. This infestation was given immediate attention due to the potential for damage in the state. Traps were placed in the park almost immediately by the Bureau of Entomology, California Department of Agriculture, which was given responsibility for organizing and carrying out an eradication program. The first step was to determine the extent of the infestation. A total of 2,800 traps were placed in an area with a radius of 25 miles surrounding Sacramento. Trapping was carried out until September, supplemented by visual inspections focusing on roses.

Foliage spray treatments covered areas within a one-half mile radius of known infestations. Initially, this included about 250 city blocks around Capitol Park, the center of the infestation. Later a larger infestation was discovered southwest of the park, leading to the treatment of an additional 50 blocks. The total area treated covered approximately 1,000 acres. Carbaryl (Sevin) was chosen as the foliage spray because of observations on its use in the eastern United States. It acted quickly against the beetle but showed little overt toxicity to plants, wildlife, and humans.

Several complaints were received from homeowners who thought the spray had damaged their flowers and shrubs. The agency determined, however, that the unusually hot weather that summer, combined with heavy spider mite populations or excessive watering, was actually to blame.

After initially using a combination of turbine and hydraulic sprayers, it was found that low-volume, high-concentration mist sprayers worked best for foliage above a height of fifteen feet. These allowed application of an adequate amount of the insecticide without leaving yards and gardens soaking wet. Later, in July, two power dusters were supplied by the city of Sacramento, and a 10 percent carbaryl dust was applied to street trees in the treatment area to minimize chances for liquid insecticide damage to automobiles parked along streets. Based on recommendations of entomologists in the past, foliage sprays were applied at ten-day intervals during the flight period of the adult beetle. These recommendations were based on the residual activity of carbaryl and the growth of plants.

Several factors favored eradication of the Japanese beetle in California. First, there were no outlying infestations that could serve to reinfest the area. The infestation was isolated and could not colonize arid surrounding land. Second, since there is no summer rainfall, only irrigated areas could support the beetle, thus limiting the treatment areas. Third, since only irrigated areas could support the beetle, the adults were not likely to fly long distances in search of food.

Critical to success of the eradication program was elimination of larval populations by soil treatment. Chlordane was chosen for this based upon its efficacy and lower risk to humans than aldrin, dieldrin, and endrin. The public was expected to readily accept its use because the chemical was commonly used in California. Soil tests done to determine the best formulation and rate of application resulted in an application rate of 10 lbs per acre in different formulations. In the summer of 1962 only five adults were captured in the treated area (Fleming 1976). The untreated areas were then also treated, resulting in a total of 3,500 treated acres. Surveys in 1963 and 1964 failed to find any beetles, and eradication was considered successful.

To evaluate the use of an ultralow volume aerial application of undiluted 95 percent technical malathion in controlling an outbreak of the Japanese beetle, a cooperative study was done by the Tennessee Department of Agriculture and the Plant Pest Control and Entomology Research Divisions of the USDA (Hamilton et al. 1967). The test area was a 400-acre valley in eastern Tennessee, where three Japanese beetle traps had captured 19 adults in 1964. Traps surrounding this valley caught no beetles. The valley was ideal for testing because its farmland, lawns, orchards, and gardens could easily support the beetle, but the valley itself was isolated by steep wooded slopes. The center of the infestation seemed to be a farm with apple trees and grapes. Evaluation of treatments were done at this location.

The insecticide was applied by helicopter seven days after the first adults were seen on June 1, 1965. This first application was at 4 oz per acre, but poor results led to the increase to 8 oz per acre. Eight applications were made at weekly intervals, until the first week of August. In 1966 adults were seen on June 8, and aerial applications were made one and two weeks later. The sprays apparently slowed the normal population buildup but did not eradicate the beetle.

It was thought that an insecticide with a longer residual time might work better. Therefore carbaryl was tested against an isolated infestation in Illinois (Hamilton et al. 1971). The test area was 8,000 acres near East St. Louis surrounding a 10-acre soybean field. The area sprayed included farmland, railroad, highways, homes, stockyards, and a chemical plant. Evaluation was done using traps. In 1966 four applications were made to the test area using 80 percent wettable powder at 1 lb/gal water per acre, applied by airplane. Between 1966 and 1967 the beetle population decreased 97 percent but increased slightly in 1968, although this still represented a decrease of 85 percent from 1966. Rainfall in the years during this study was below normal, especially in 1966 and 1968. This probably contributed to control of the beetle, although the carbaryl treatments were thought to have been primarily responsible.

San Diego Study

The Japanese beetle invaded San Diego County, in California, in 1973. An area comprising 100 residential blocks in San Diego was declared an eradication zone. The beetle was successfully eradicated by 1974. A study was undertaken to document the effect of the pesticides used for the eradication of the Japanese beetle on the natural enemies of the major citrus pests: woolly whitefly (*Aleurothrixus floccosus* [Maskell]), purple scale (*Cornuaspis beckii* [Newman]), citrus mealy bug (*Planococcus citri* [Risso]), and citrus red mite (*Panonychus citri* [McGregor]) (DeBach and Rose 1977).

In 1971 parasites of the woolly whitefly had been introduced, and biological control was achieved by 1973. The other citrus pests were also under good biological control. In December 1973, three months after initial foliar carbaryl treatment and ground application of chlordane, no live parasites of the woolly whitefly could be found in the eradication area, while outside the zone the parasites were abundant. The woolly whitefly populations did not, however, increase as expected. Rather, the citrus red mite population increased so rapidly that acute defoliation occurred with over 80 percent of the study trees affected. Carbaryl is believed to have physiological effects on citrus that may make the foliage a more suitable habitat for citrus red mite. The defoliation eliminated a substantial part of the woolly whitefly population, which occurs on leaves. At the same time, purple scale and citrus mealy bug populations increased.

During the second eradication season in the summer and fall of 1974 there was again severe defoliation with 90 percent of the treated trees affected. Woolly whitefly increased dramatically on trees without severe defoliation. Purple scale continued to increase as its major natural enemies were largely destroyed by carbaryl. A single dicofol application was made to control the citrus red mite. Dicofol, however, is 50 times more harmful to predaceous mites than to citrus red mite. By September and October 1974, citrus red mite was devastating the eradication area.

Carbaryl treatments ended in September 1974. Both woolly whitefly and purple scale, freed of biological and chemical restraints, began to increase rapidly. Defoliation caused by citrus red mite soon followed and reduced the woolly whitefly population. Purple scale living primarily on the woody portions of the trees was not as affected. By December 1974, some citrus trees began to die from the combined stresses caused by the three major pests—woolly whitefly, purple scale, and citrus red mite—and by May–June 1975 some trees had been completely destroyed. By October 1975, natural enemies of woolly whitefly had become reestablished but had not yet achieved biological control. At this time the woolly whitefly population in the eradication zone was twelve times greater than in untreated areas.

Before the eradication program purple scale was generally distributed but harmless to citrus. By October 1974, purple scale had increased so that more than 60 percent of the study trees showed injury. By September 1975, one year after treatment with carbaryl, the ratio of live scales in the eradication versus the control zone was 12,000 to 1. This study supports previous findings that insect eradication programs may cause environmental disruptions.

Evaluation

The Japanese beetle is firmly established in the eastern United States and will probably continue to spread naturally until it is stopped by environmental restrictions, primarily temperature and rainfall. The infestation will also serve as a source for isolated colonies in other areas. Therefore, eradication efforts will continue in the future.

As in the past, many factors will contribute to the degree of success of these projects. Most successful eradication efforts were against small isolated populations treated soon after detection. This early detection is probably the most important factor in limiting the spread of the beetle. Eradication usually involved extensive use of insecticides, in one case up to 100 lbs of lead arsenate per acre per year for ten years. In some areas eradication was probably due more to climatic factors than to any efforts by humans. In other areas, such as Sacramento, a combination of environmental resistance and chemical treatment seem to have been responsible for the success of the eradication projects.

Eradication efforts can have a severe impact on the environment, as demonstrated by the San Diego study. The ecological and economic consequences of any eradication effort should be carefully considered before the attempt is carried out.

Literature Cited

Commonwealth Institute of Entomology. 1978. *Distribution maps of insect pests*. Map no. 16. London: Commonwealth Institute of Entomology.

Dawson, J. C. 1936. Japanese beetle in the middle west. *J. Econ. Entomol.* 29:778–80.

———. 1937. Progress of Japanese beetle suppression in St. Louis. *J. Econ. Entomol.* 30:611–14.

———. 1938. The present status of the Japanese beetle situation in St. Louis. *J. Econ. Entomol.* 31:590.

DeBach, P., and M. Rose. 1977. Environmental upsets caused by chemical eradication. *Calif. Agric.* 31:8–10.

Dunbar, D. M., and R. L. Beard. 1975. Present status of milky disease of Japanese and Oriental beetles in Connecticut. *J. Econ. Entomol.* 68:453–57.

Fleming, W. E. 1958. Biological control of the Japanese beetle especially with entomogenous diseases. *10th Internat. Cong. Entomol. Proc. 1956* (3):115–25.

———. 1963. The Japanese beetle in the United States. USDA *handbook* 236.

———. 1968. Biological control of the Japanese beetle. USDA *tech. bull.* 1383.

———. 1972. Biology of the Japanese beetle. USDA *tech. bull.* 1449.

———. 1976. Integrating control of the Japanese beetle: A history review. USDA *tech. bull.* 1545.

Fleming, W. E., and I. M. Hawley. 1950. A large-scale test with DDT to control the Japanese beetle. *J. Eco*. *Entomol.* 43:586–90.

Fox, H. 1927. The present range of the Japanese beetle, *Popillia japonica* Newm., in America and some factors influencing its spread. *J. Econ. Entomol.* 25:396–407.

———. 1932. The known distribution of the Japanese beetle in 1930 and 1931, with special reference to the area of continuous infestation. *J. Econ. Entomol.* 25:396–407.

———. 1934. The known distribution of the Japanese beetle in 1932 and 1933. *J. Econ. Entomol.* 27:461–73.

———. 1939. The probable future distribution of the Japanese beetle in North America. *J. N.Y. Entomol. Soc.* 47:105–23.

Gammon, E. T. 1961. The Japanese beetle in Sacramento. *Calif. Dept. Agric. Bull.* 50:221–35.

Goonewardene, H. F., J. E. McKay, D. B. Zepp, and A. E. Grosvenor. 1970. Virgin female beetles as lures in field traps. *J. Econ. Entomol.* 63:1001–3.

Hadley, C. H., and I. M. Hawley. 1934. General information about the Japanese beetle in the United States. USDA *Circ.* 332.

Hamilton, D. W., W. H. Luckman, M. A. Campbell, W. W. Maines, R. W. Bills, and L. B. Matzenbacher. 1971. Low-volume-aerial application of carbaryl over a large area to reduce a population of Japanese beetles. *J. Econ. Entomol.* 64:68–70.

Hamilton, D. W., W. W. Maines, A. J. Coppinger, and H. L. Bruer. 1967. Ultra-low-volume-technical malathion for suppression of an incipient infestation of Japanese beetle. *J. Econ. Entomol.* 60:1480–81.

Hawley, I. M. 1944. Notes on the biology of the Japanese beetle. *U.S. bureau of entomology and plant quarantine*, E–615.

———. 1949. The effect of summer rainfall on Japanese beetle populations. *J. N.Y. Entomol. Soc.* 57:167–76.

Hawley, I. M., and T. N. Dobbins. 1945. The distribution and abundance of the Japanese beetle from 1935 through 1943, with a discussion of some of the known factors that influence its behavior. *J. N.Y. Entomol. Soc.* 53:1–20.

Hawley, I. M., and F. W. Metzger. 1940. Feeding habits of the Japanese beetle. USDA *Circ.* 547.

Hutton, P. O., and P. O. Burbutis. 1974. Milky disease and Japanese beetle in Delaware. *J. Econ. Entomol.* 67(2):247–48.

Klein, M. G., W. R. Nickle, P. R. Benedict, and D. M. Dunbar. 1976. *Psammomermis* sp. (Nematoda: Mermithidae): A new nematode of the Japanese beetle, *Popillia japonica* (Coleoptera: Scarabaeidae). *Proc. Helminthol. Soc. Wash.* 43(2):235–36.

Ladd, T. L., Jr. 1970. Sex attraction in the Japanese beetle. *J. Econ. Entomol.* 63:905–8.

Langford, G. S., F. B. Washington, R. H. Vincent, and E. N. Cory. 1941. Cooperative Japanese beetle work in Maryland. *J. Econ. Entomol.* 34:416–18.

Lawrence, K. O. 1981. Japanese beetle: Control of larvae with isofenphos. *J. Econ. Entomol.* 74(5):543–45.

Luckmann, W. H., and G. C. Decker. 1960. A 5-year report of observations in the Japanese beetle control area at Sheldon, Illinois. *J. Econ. Entomol.* 53:821–27.

Major, R. T., and H. J. Tietz. 1962. Modification of the resistance of *Gingko biloba* leaves to attack by Japanese beetles. *J. Econ. Entomol.* 55:272.

Regniere, J., R. L. Rabb, and R. E. Stinner. 1979. *Popillia japonica* (Coleoptera: Scarabaeidae): A mathematical model of oviposition in heterogeneous agrosystems. *Can. Entomol.* 111;1271–80.

———. 1981a. *Popillia japonica:* The effect of soil moisture and texture on survival and development of eggs and first instar grubs. *Environ. Entomol.* 10:654–60.

———. 1981b. *Popillia japonica:* Intraspecific competition among grubs. *Environ. Entomol.* 10:661–62.

———. 1983. *Popillia japonica* (Coleoptera: Scarabaeidae): Distribution and movement of adults in heterogeneous environments. *Can. Entomol.* 115:287–94.

Sharpe, E. S. 1976. Toxicity of the parasporal crystal of *Bacillus thuringiensis* to Japanese beetle larvae, *Popillia japonica. J. Invert. Path.* 27:421–22.

Sharpe, E. S., and R. Detroy. 1979. Susceptibility of Japanese beetle, *Popillia japonica* larvae to *Bacillus thuringiensis:* Associated effects of diapause, midgut pH, hydrogen-ion-concentration and milky disease. *J. Invert. Path.* 34:90–91.

Smith, L. B. 1924. The Japanese beetle status in 1923. *J. Econ. Entomol.* 17:107–11.

Smith, L. B., and C. H. Hadley. 1926. The Japanese beetle. USDA *Circ.* 363.

Stockwell, C. W. 1935. The Japanese beetle outbreak in St. Louis, Mo., and its control. *J. Econ. Entomol.* 28:535–37.

Tashiro, H., J. B. Bourke, and S. D. Gibbs. 1981. Residual activity of dieldrin and chlordane in soil of established turf in Japanese beetle grub control. *J. Econ. Entomol.* 74:397–99.

U.S. Department of Agriculture (USDA). 1973. Controlling the Japanese beetle. *Home and Garden Bull.* 159. Washington D.C.: Government Printing Office.

van Leeuwen, E. R. 1932. Reactions of the Japanese beetle to spray deposits on foliage. USDA *Circ.* 227.

8

Whitefringed Beetle

Michael J. Pitcairn and Stephen A. Manweiler

Since its detection in Okaloosa County, Florida, in 1936, the whitefringed beetle, *Graphognathus* spp. (Coleoptera: Curculionidae), has proved to be a destructive pest of many field and garden crops throughout the southeastern United States. Serious damage is caused by larvae feeding on the roots of plants. These beetles are able to exist under varied climatic conditions, and adults and larvae are able to feed on more than 300 species of plants (USDA 1972).

Severe damage has been reported to occur in cotton, corn, peanuts, potatoes, and many truck and garden crops. In 1937 a state quarantine of the infested acreage was established and was followed by a federal quarantine in 1939. The objectives were to suppress established populations, eradicate new outlying infestations, and restrict spread of the beetles through commercial channels.

Various control methods were attempted, but the most effective was the incorporation of chlorinated hydrocarbon insecticides, especially DDT, into the soil. These substances had relatively long residual toxicities and provided control for three to five years without repeated applications. By 1971 whitefringed beetles had spread to thirteen states in the southeast with the heaviest infestations occurring in Georgia, Florida, and Alabama. In 1975 the federal whitefringed beetle quarantine was revoked; responsibility for its control currently depends on individual growers and transporters.

Life History and Habits

In the spring of 1936 damage to a peanut field near Svea (Okaloosa Co.), Florida, caused by numerous white grubs was reported to a local agricultural agent (Young, App, and Green 1938). Adult beetles collected that summer by the county agent were identified by L. L. Buchanan as *Pantomorus leucoloma,* the whitefringed beetle (Warner 1975). The beetle was originally described by Boheman in 1840 as *Naupactus leucoloma* from Argentinian specimens, and the number of valid species actually occurring in the United States has been the subject of considerable disagreement. Buchanan (1939) described the subgenus *Graphognathus* and included six U.S. species: *Pantomorus (Graphognathus) leucoloma, P. (G.) peregrinus, P. (G.) minor, P. (G.) pilosus, P. (G.) striatus,* and *P. (G.) dubius* (Buchanan 1942). In 1947 he published a correction stating that his 1936 identification of material from Svea as *Pantomorus leucoloma* was wrong. This he described as a new species, *Pantomorus (Graphognathus) fecundus; P. leucoloma* was considered not to be present in the United States (Buchanan 1947). Warner (1975) raised *Graphognathus* to generic rank and synonymized three of Buchanan's species (*striatus, dubius,* and *pilosus*) with *Graphognathus leucoloma.* Thus *G. leucoloma* does occur in the United States. Warner (1975) considered *G. fecundus, G. minor,* and *G. peregrinus* to be valid species.

The biologies of all four *Graphognathus* species were similar enough that the control measures developed for any one species worked equally well against the others. The importance of precise species identification becomes more acute, however, as the control measures become more species-specific. The following is a general description of their life history and of the damage caused by this species complex. Specific differences are not discussed.

The whitefringed beetle (WFB) is univoltine and spends all but about one month of the year in the soil (fig. 8.1) USDA 1962a). The adults, about 1.1 cm long, are dark gray with a whitish band along the dorsolateral margin of the body. They cannot fly because their elytra are fused together along their inner margins. They may, however, migrate by walking, possibly as much as three-quarters of a mile in a life span (Young, App, and Green 1938). Adults begin emerging from the soil in early May and continue emerging through mid-August (Henderson and Padget 1949). All adults are parthenogenic

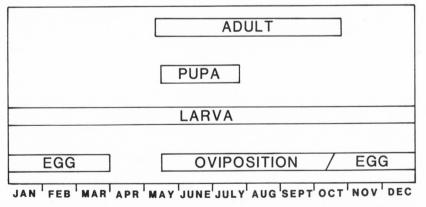

Figure 8.1. Seasonal cycle of the whitefringed beetle, *Graphognathus* spp. Bars identify time periods when oviposition and the four life stages are usually present.

females. Egg laying begins five to twenty-five days after emergence (Young et al. 1950). Eggs are laid in masses covered with a sticky substance and are usually located where objects such as plant stems, sticks, gravel, or garden tools contact the soil (fig. 8.2). Most masses contain fifteen to twenty eggs, although sixty eggs per mass has been observed. Oviposition occurs from late May to mid-October (USDA 1962a).

The adults live about three months and oviposit intermittently through-out their lives. Fecundity depends upon the food source available to the adult beetle and ranges from very high in adults feeding on some one hundred and seventy species of field plants to an average of only four eggs per adult among those feeding on grass (Young et al. 1950).

Proper temperature and moisture conditions are required for hatching the eggs, which are yellowish and about 0.09 mm long. During summer the eggs hatch in about seventeen days but may require over a hundred days if

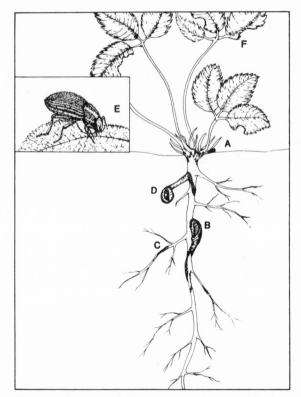

Figure 8.2. Life stages and damage caused by whitefringed beetle. A = egg mass; B, C = larvae; D = pupae, E = adult and damage (6x); F = adult damage (adapted from USDA 1972).

laid in December. Some remain viable throughout the winter and hatch the following spring (ibid).

The larvae inhabit the upper nine inches of soil near, or in contact with, the roots of food plants (fig. 8.2). Soon after hatching they tunnel to these roots and remain until the roots become unsuitable food or until pupation occurs. Feeding begins at emergence and continues until cold weather arrives. Feeding resumes with the warmer spring temperatures. The beetles require about eleven months to complete their development but can take as much as two years depending upon environmental conditions (Metcalf, Flint, and Metcalf 1962).

Fully mature larvae, about 1.25 cm long, pupate within an earthen cell usually two to six inches deep from early May through July (fig. 8.2). They do not move long distances from host plant roots before pupation. The pupal stage lasts about eight to fifteen days, after which the adult may remain within its cell for a few days while its exoskeleton hardens. It then burrows its way to the surface. Emergence frequently occurs after rain (Henderson and Padget 1949).

The larvae are polyphagous. They have been observed feeding on over 385 species of plants including carrots, Irish potatoes, sweet potatoes, peanuts, corn, sugarcane, cotton, cowpeas, velvetbeans, cabbage, collards, chufa, alfalfa, Mexican clover, blackberries, chrysanthemums, clover, oats, peas, strawberries, soybeans, and tobacco. Wild hosts include carelessweed, cocklebur, curleydock, lupine, mustard, and ragweed. The roots of young woody plants such as peach, pecan, tung, willow, and pine seedlings are eaten. Larvae have been found attacking the swollen seeds of acorns, peachseeds, pecans, and tungnuts (Henderson and Padget 1949; Metcalf, Flint, and Metcalf 1962; Young et al. 1950; USDA 1962a).

The major damage caused by WFB on agricultural crops is due to root feeding by the larvae (fig. 8.2). They cause more damage to plants with main (tap) roots than those with fibrous roots because their feeding can completely sever the root. They also burrow into the tubers and enlarged roots of carrots, potatoes, and turnips. Although the plant may not be killed, its market value is destroyed (USDA 1962a). Larval feeding may cause a plant to wilt and die, and surviving plants are usually stunted or malformed. Damage is especially bad during the spring when the larvae are concentrated in the upper few inches of soil and the majority are reaching maturity. A few cases of severe damage to fall or winter crops have occurred, but usually feeding is limited at this time of year (Henderson and Padget 1949). In some fields up to 70 percent of the plants have been killed in areas ranging from a few square yards to twenty acres. Frequently a destroyed field will be replanted, only to be destroyed again; eventually, the land may be abandoned (Young et al. 1950).

Adult WFB cause damage by feeding on the leaves of over more than a hundred and seventy species of plants. This damage is usually less severe than that caused by the larvae. Food plants include field, garden, and truck crops, weeds, ornamental shrubs, flowers, wild bushes, vines, and trees. The beetles prefer plants with broad leaves, and those with smooth leaves are often chosen over those with hairy leaves (Young et al. 1950). Feeding is done at the leaf edge, making "sawtooth cuts" (fig. 8.2) rather than holes (Henderson and Padget 1949). Only rarely is this damage severe enough to affect crop yields significantly. Sometimes, when the beetles are abundant, defoliation occurs (Young et al. 1950). A population of 240,000 beetles per acre was estimated for one cotton field that suffered severe damage. Two hundred and fifty beetles were collected from one plant (Henderson and Padget 1949). On two-month-old soybean plants, populations of fifty or more adult beetles per plant caused significant reductions in yield (Ottens and Todd 1980). But the occurrence of these adult abundances in the field is very unusual. In spite of their documented host range, adults appear to have definite food preferences for peanuts, velvetbeans, soybeans, lespedeza, and alfalfa.

In many areas damage was most severe from 1937–42. Near Florala, Alabama, several hundred acres of field crops were completely destroyed. In 1938 more than 1000 acres were left unplanted due to high larval populations (Young et al. 1950). More recently, Boutwell and Watson (1978) reported that in 1975 losses in peanuts due to the whitefringed beetle (larval and adult damage) were estimated to be $3.6 million (Alabama), $1.1 million (Florida), and $8.4 million (Georgia), giving a total of $13.1 million for these three states alone.

Epidemiology and Ecology

The WFB originated in South America, where it is endemic in Uruguay, Chile, Argentina, and southern Brazil. Infestations have also occurred in New South Wales, Australia, the Union of South Africa, and New Zealand (East 1980; Henderson and Padget 1949) (fig. 8.3). After the initial discovery in 1936 of *Graphognathus* in the United States in Okaloosa County, Florida, it was soon found in Walton County, Florida, and in two counties in Alabama as well. In 1937 inspections of ports and areas of commerce revealed additional infestations in Florida, in three countries in Mississippi, and in New Orleans, Louisiana (Young, App, and Green 1938). It is possible that the introduction of *Graphognathus* occurred through the dumping of ballast from ships engaged in commerce with South American countries (Sailer 1983).

By 1938 WFB infestations were located in twenty-one counties in four states; by 1944 140,000 acres in five states were affected. Whitefringed

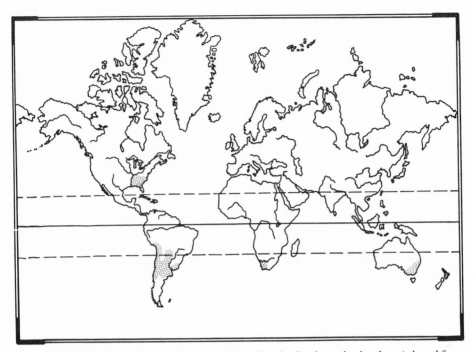

Figure 8.3. Worldwide distribution of the whitefringed beetle, *Graphognathus leucoloma* (adapted from Commonwealth Institute of Entomology 1964).

beetle was reported for the first time in Georgia in 1946. The Georgia infestations were not limited to agricultural land but were also found on residential and industrial sites and in large ornamental nurseries, especially in the municipalities of Eastman, Fort Valley, and Macon. A follow-up inspection of properties landscaped with plants obtained from infested nurseries disclosed many additional infestations in Georgia, South Carolina, and Alabama. This follow-up inspection of landscaping projects developed into a new tactic of the WFB survey. Identifying nursery articles capable of harboring the pest and tracking their movements to new landscape projects became an effective way to discover additional infestations (USDA 1947).

The WFB continued to spread; by 1948 220,000 acres were considered to be infested throughout the southeastern United States. In 1952 WFB was found in one hundred and fourteen counties in eight states, although ten counties were free of the beetles for three consecutive years and were taken from the restricted list. By 1960 more than one million acres were reported infested in eleven states; by 1980 WFB had spread to thirteen states (fig. 8.4).

Information on the spread of *Graphognathus* is indicated by the number

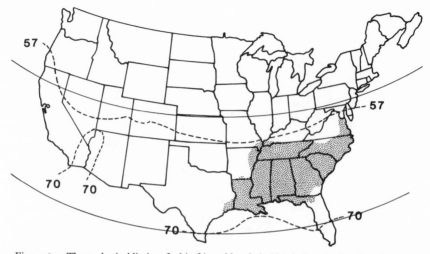

Figure 8.4. The ecological limits of whitefringed beetle in North America based on known
limits in South America. The solid lines indicate the corresponding latitudes between
which whitefringed beetle is known to exist in South America. The dotted lines indicate
the 70° F isotherm and 57° F isotherm (adapted from Henderson and Padget 1949).

of counties reporting new infestations each year in the annual report of
the Administrator of Agricultural Research for the years 1936–50, from the
Cooperative Economic Insect Report for the years 1950–75, and from the
Cooperative Plant Pest Report for the years 1976–80. There were four
periods of dissemination. The first period of dissemination, occurring prior
to 1940, was probably due to the intensive survey activities that were performed and not due to actual invasion and spread of WFB into new areas.
The second period of dissemination occurred from 1940–44 and the third
increase in dispersal occurred during the years 1945–50. Last, a major period
of dispersal began in 1960 and lasted until 1973.

The quarantine of WFB was terminated in 1975 because the pest was
thought to have reached the limits of its ecological range within the eastern
United States. The potential range of WFB was determined from its distribution in its native home in South America. On the South American
continent, the southern limit of WFB is around 40° south latitude, and the
northern limit occurs between 23° and 29° south latitude. In the northern
hemisphere these correspond to the latitude of Philadelphia, Pennsylvania
in the north and the southern region of the Florida peninsula in the south,
respectively. In South America, the southern and northern limits may also
be approximated by the mean annual temperature isotherms of 57°F and
70°F, respectively (Henderson and Padget 1949). Figure 8.4 illustrates
where these limits are located in North America. Given this evidence it does
appear that the WFB has expanded to much of its potential range in the

eastern United States. California is, however, a good candidate for the next area of introduction and spread.

Because the beetle cannot fly and natural dissemination is very slow, it is probable that much of its dispersal has come through commerce. The following factors in the biology of the beetle allowed them to be easily transported through commercial channels: (1) Eggs deposited on plant stems, sticks, gravel, and other objects on the ground will hatch only under favorable conditions (for example, high moisture) but may remain viable for more than seven months. Hay from infested fields and infested soil attached to nursery stock are good means of transport (Young et al. 1950). In addition, adults conceal themselves when disturbed and hang tenaciously onto seed cotton, hay, and other crops, as well as onto vehicles, agricultural equipment, and even humus. (2) Their cosmopolitan feeding habits allow widespread establishment and increase the number of possible carrier materials. (3) The parthenogenic mode of reproduction allows a new infestation to be initiated by any stage in the life cycle. Once established, populations build up very quickly.

Control and Eradication History

After the initial discovery of the WFB in 1937, preliminary investigations of its economic impact showed that it could be a serious pest on a wide range of cultivated crops. This realization resulted in a special allotment to the Bureau of Entomology and Plant Quarantine, USDA (hereafter referred to as BEPQ) for control and suppression of the WFB. More than sixty federal and state inspectors and four hundred laborers were eventually enlisted in this initial campaign (Young, App, and Green 1938).

First an intensive survey was made of the known infestation area in southern Alabama and western Florida. Once this area was delineated, other inspections were made at numerous ports and commercial areas in Texas, Louisiana, Mississippi, Alabama, Florida, and Georgia. Additional infestations were found at Pensacola, Florida, in Laurel, Mississippi, at New Orleans, Louisiana, and in Harrison and Stone Counties, Mississippi (Young, App, and Green 1938).

Subsequent to this, all infested areas in Alabama, Mississippi, and Florida were placed under rigid quarantines. The primary objective of this early state cooperative control project was to restrict intrastate movement of the insect and to reduce its numbers by suitable control tactics. During that same year a research project in the BEPQ Division of Cereal and Forage Insect Investigations at Florala, Alabama, was established to study the biology of WFB (Young et al. 1950).

To prevent its spread into new territories a barrier ditch was constructed around each infested field. An outside barrier ditch approximately 150 miles long was constructed around the entire infestation in the Florala area to prevent the beetle from walking to new areas. In addition, grasses and weeds were removed from areas surrounding lumberyards, sawmills, cotton gins, railway yards, rights-of-way, and other industrial sites. As a further precaution, the municipalities of Laurel, Mississippi, and Pensacola, Florida, cooperated by clearing vacant lots and alleys. Weed burners were used for heavy vegetative cover (Young, App, and Green 1938).

The objective of this early phase of quarantine efforts was to prohibit further spread rather than to control WFB. As more information about the biology of the beetle became known, quarantine procedures were revised accordingly.

On January 15, 1939, a Federal Quarantine (#72) was established to restrict interstate movement or activities believed capable of disseminating the beetles (USDA 1938). This mainly restricted the movement of certain soil and plant materials. Articles leaving the regulated area were inspected and, if not free of the beetle, were fumigated with methyl bromide (McClurkin 1953).

In addition to establishing a federal quarantine, the BEPQ also entered into a cooperative control project with the states of Alabama, Louisiana, and Mississippi. This federal-state control project had the following objectives: "(1) prevention of dissemination of beetles by drastic suppression of population and by regulatory measures; (2) drastic suppression aimed at eradication of isolated infestations; and (3) protection of crops by furnishing materials and equipment to farmers" (USDA 1943, 165).

During the early years of the cooperative control project (1939–45) the insecticides calcium arsenate and cryolite were used. Calcium arsenate was the first insecticide actually used extensively against WFB. Because this chemical burns the foliage of leguminous plants (for example, peanuts and velvetbeans) and because it was extremely toxic to livestock, its use was superceded by cryolite. Yet, even though high adult mortalities were obtained with calcium arsenate and cryolite, larval densities were not substantially reduced. Very high adult mortalities appeared to be necessary to overcome the insect's high reproductive potential and effectively reduce future larval populations (Henderson and Padget 1949; Young 1960).

In addition to the population suppression activities, the WFB control project also conducted extensive surveys throughout the noninfested areas. Many survey and inspection methods in use today such as the tracing of nursery materials were originally developed for this control project (Rohwer 1953, personal communication, 1983).

Prior to the introduction of the WFB, a common farming practice in the southeastern United States was to intercrop velvetbeans with corn to provide a winter grazing crop for cattle. Once WFB became established, this practice was no longer practical. Velvetbean was a preferred host plant, and in spite of control measures beetle numbers usually became very high whenever this crop was grown. Corn is not a preferred host but would suffer much damage because of its close proximity to the velvetbeans. In response to this, intercropping was discontinued, and monocultures were rotated on a three or four year cycle. The economic impact of this beetle was most obvious in the major changes of farming methods that were practiced following its establishment (G. Rohwer, personal communication, 1983).

Using a rotation plan, crops preferred by WFB, such as peanuts, soybeans, and velvetbeans, could still be successfully grown. The objective of the rotation scheme was to prevent the primary adult food plants from being grown more than once every three or four years on the same land. The addition of organic matter to the soil also caused a reduction in damage (Smith 1948). The use of cryolite insecticide sprays and crop rotation were the primary methods of WFB control during the early forties.

Immediately following World War II, the insecticide DDT was tested and shown to be a very effective adulticide and larvicide against WFB (Young 1944). In addition it gave good larval control for three to five years following a single application. Not surprisingly, DDT became the primary tool for WFB control. The failure of earlier suppression efforts was their inability to achieve any significant larval mortality. Later other chlorinated hydrocarbon insecticides were developed and were also found to give good larval control (Young and Gill 1948; Young 1960) (table 8.1).

The most effective method of larval control was to incorporate the insecticide into the soil. DDT was incorporated into fertilizers that were applied at planting, and the convenience of this method made it the one most preferred. Drill row treatments were also used but had to be applied annually for three to four years before residual insecticidal effects were achieved (Young et al. 1950; Young 1960).

To prevent further dissemination, noncultivated lands such as lawns, industrial and railroad properties, roadsides, and wastelands were also treated. In the cooperative federal-state program, repeated foliage applications were required each year on noncultivated lands. Tests begun in 1950 showed that dieldrin gave the best results. Surface treatments with dieldrin at 3–5 pounds active ingredient per acre eventually replaced the previous foliage treatments using cryolite or DDT (Young 1960).

To allow transfer of materials in and out of the quarantine area the use of DDT as an ovicide was developed. One pound active ingredient in 7.5 gallons of kerosene proved effective. In cases where this ovicide could not be used because of its oil base, benzene hexachloride in water was developed. This was applied in a concentration of one pound gamma isomer of benzene

Table 8.1 Recommended Dosages of Various Insecticides to Control the Whitefringed Beetle during the 1950s and Early 1960s

Type of Land and Treatment	Insecticide[1]	Type of Application			
		Broadcast		Drill Row[2]	
		Pounds per Acre	Years Effective	Pounds per Acre	Years Effective
Cultivated (soil treatment)	Aldrin	2	≥3	0.75–1	1
	Chlordane	5	3	1–2	1
	DDT	10	4	2–3	1
	Dieldrin	1.5	≥4	0.5–0.75	1
	Heptachlor	2	≥3	0.75–1	1
Noncultivated (surface treatment)[3]	Dieldrin	3	≥4	—	—

Source: Young (1960).

[1]For use only on crops on which they are currently approved.
[2]Deposit insecticide below depth of seed. If the distance between drills is less than 3.5 feet, apply the maximum dosage; if it is more than 6 feet, apply the minimum dosage. Row treatment is not recommended on vegetable crops.
[3]Application should be made only under general direction of federal or state plant pest control personnel.

hexachloride in 100 gallons water and eventually superceded the DDT treatment as the most commonly used compound to control WFB eggs (Young 1960).

With the introduction of DDT in 1946, a big push for eradication was initiated by the federal-state cooperative control project. Soil treatments of 10 pounds active ingredient of DDT per acre on cultivated farm land were performed and financed by the BEPQ. Farmers were urged to apply supplementary treatments of DDT mixed with commercial fertilizers. The objective of following up the initial soil treatment with annual applications of the DDT-fertilizer mix was to build residues of DDT high enough to prevent reinvasion of treated fields (USDA 1951).

Between 1946 and 1952 more than 52,000 acres of agricultural land had been soil-treated with 10 pounds active ingredient of DDT per acre. Between 1948 and 1952, about 3,100 acres of nursery land had received soil treatments of 50 pounds active ingredient DDT per acre. During 1951 and 1952, some 6,000 acres of industrial and other noncultivated land received a surface application of 25 pounds active ingredient per acre (USDA 1953a).

By the end of 1952 WFB had spread to a total of one hundred and fifty-two counties in eight states. However, the control efforts outlined above were not totally without merit. The rate of spread during the early 1950s was significantly reduced from the earlier period of 1946–48. The USDA annual report for the fiscal year 1952 (USDA 1953b) reported that while nine additional counties were found to be infested for the first time, ten other counties (seven in Georgia; one in Mississippi; one in North Carolina; and one in South Carolina) were found free of the beetles for three consecutive years and removed from the infested list. Of the one hundred and forty-two counties recognized as having active infestations, "376,755 acres [were] classified as infested. No specimens could be recovered during 1952 on approximately 24 percent of the acreage classified as infested; light populations were present on 48 percent of the infested acreage; moderate populations on 24 percent; heavy populations on only 4 percent. . . . The beetle has not been reported in a new State since 1948 even though extensive surveys have been made" (USDA 1953b, 328).

In September 1951 Charles F. Brannan, secretary of agriculture, appointed an independent study group to review and evaluate the insect and plant disease control programs of the BEPQ. In their report submitted February 1, 1952, the following recommendations were made: (1) Eradication of the WFB with available control measures was not considered practical; (2) The control program should be continued to prevent spread and reduce damage by the beetle; (3) Now that effective control measures have been developed, including specialized application equipment, control of cultivated lands should be executed by the farmers or landowners. It was recommended that federal assistance of control methods on farm lands be discontinued; (4)

Research, survey, and technical assistance and educational programs should continue to be provided (Minor et al. 1952).

The shift of the financial burden of the WFB control program is most evident in the 1952 annual report of the BEPQ:

Since 1946 more than 52,000 acres of agricultural lands have been soil-treated with 10 pounds of DDT per acre for whitefringed beetle control. Farmers and growers have now assumed this phase of control work. Since 1950 farmers have applied DDT-fertilizer mixtures to an additional 10,000 acres of farmland. Since 1948 about 3,100 acres of nursery land have been soil-treated with 50 pounds of DDT per acre. Responsibility for this procedure, which allows certification of nursery plants for movement without further treatment, has also been assumed entirely by the growers. During 1951 and 1952, some 6,000 acres of industrial and other noncultivated land, on which DDT could not be cultivated into the soil, received a 25-pound per acre surface application.

During the year expenditures by the states, counties, municipalities, industrial concerns, growers, and others exceeded the expenditure by the Bureau. Insecticides were furnished and applied in the greater part by the individual concerned. Some insecticides were supplied by the states. (USDA 1953b, 83)

In spite of the shift of financial burden from USDA to landowners, there was little appreciable change in the objectives of the federal-state cooperative control project concerning eradication and suppression of WFB. The federal-state cooperative control activities from 1946 to its termination in 1975 can be summarized as follows: (1) treatment of as much of the currently infested area as possible with a soil application of DDT at 10 pounds active ingredient per acre. Each year, annual reports of the BEPQ reported the cumulative total acres treated since 1946. These treatments were supplemented by annual applications of a fertilizer-DDT mixture by individual growers in the attempt to build up soil residues to greatly suppress the beetle populations and reduce the likelihood of spread. (2) New infestations occurring outside the regulated area were hit with eradication treatments. These usually consisted of DDT at 25–50 pounds active ingredient per acre. The intention here was to eradicate these small outlying populations.

After 1950 dieldrin, applied at a rate of 3 to 5 pounds active ingredient per acre, was used to retreat beetle populations on noncultivated land (Young 1960). In 1958 eradication treatments were reported completed in one county in Alabama and in five counties in Florida (USDA 1958). In 1959 treatments of all known infested areas in South Carolina were completed (USDA 1959a) and treatments of all five infested parishes in Louisiana were completed (USDA 1959b). In 1960, eradication treatments were applied in the two counties newly infested in Arkansas. In addition, 1,039,813 acres were reported as infested in eleven states, and WFB was eliminated from 341,057 acres (USDA 1961).

The New Jersey infestation, which was discovered in 1954, was declared eradicated in 1960—this being the third consecutive year of no WFB recoveries. The Cooperative Economic Insect Report in 1961 included the following comment: "The aggressive treatment program in support of regulatory work that is followed in all areas where whitefringed beetle is found has continued to keep these pests depressed to noneconomic levels. Plant Pest Control Division and state personnel have been very successful in obtaining the cooperation of property owners to assist with eradication or control where infestations have been found" (USDA 1961, 138). In 1961 eradication treatments were applied in Virginia, Kentucky, and Arkansas (USDA 1962b).

In 1962 aerial treatments of infested acreage in two counties in Arkansas and in six counties in Tennessee were performed (USDA 1963a, 1963c). Since the beginning of the WFB program, 1.3 million acres have been found infested. More than 398,000 acres have been subjected to eradication treatments (54,612 acres were treated in 1962 alone). "Persistent control and eradication treatments have virtually eliminated economic losses from these pests." (USDA 1963b). In 1963 thirty-one new counties were reported infested with WFB. Treatments were completed on 60,000 acres (USDA 1964).

What was not known at the time these optimistic reports were written was that control of WFB was beginning to break down. Beginning in 1960 the number of new counties reported infested in the Cooperative Economic Insect Report increased dramatically. This phase of dissemination was much more extreme than what had occurred previously and ensued in spite of the

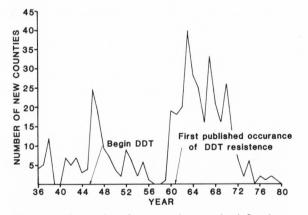

Figure 8.5. The number of new counties reporting infestations of whitefringed beetle each year (Sources: Annual Report of the Administrator of Agricultural Research for 1936–50; Cooperative Economic Insect Report for 1950–75; Cooperative Plant Pest Report for 1976–80, and Young 1960).

previous oil treatments and control tactics (fig. 8.5). In 1963, WFB larvae were found in a nursery near Semmes, Alabama, that had been surface treated with dieldrin every four years since 1954 (Harlan et al. 1972). Tests of susceptibility conducted on these larvae showed them to be resistant to dieldrin. Tests of several other infestations showed resistance of *Graphognathus* larvae to dieldrin, DDT, and heptachlor (ibid). In 1970 tests of susceptibility of WFB larvae to dieldrin sampled from nine different states indicated the presence of resistant larvae in Florida, Georgia, Tennessee, and Alabama, with the greatest resistance occurring at Semmes, Alabama (ibid).

The large increase in dissemination that occurred in the 1960s was most likely caused by the development of resistance to the commonly used chlorinated hydrocarbon insecticides, which resulted in a breakdown of the ability to suppress beetle populations. Many of the infestations first found resistant occurred in nurseries. Unfortunately, nursery stock and soil were an effective method of dissemination, and this likely contributed to its rapid spread.

A change of tactics was necessary if control of WFB was to be achieved. By the early 1970s use of DDT was being phased out, and use of dieldrin and chlordane became much more common. In 1970 eradication treatments were still performed on infestations outside the regulated area. These treatments consisted of an application of either dieldrin at 3 pounds active ingredient per acre or chlordane at 5 pounds active ingredient per acre supplemented with carbaryl (Sevin 80–S) foliar sprays applied at 0.8–1.0 pound active ingredient per acre (USDA 1970). At nurseries that were infested with WFB resistant to chlorinated hydrocarbon insecticides, eradication treatments were applied using the soil fumigant D–D at a rate of 100 gallons per acre (USDA 1970). In 1971 chlordane or dieldrin was applied to 3,798 acres of infested areas outside the regulated area. Supplemental treatments of carbaryl foliar sprays were also used (USDA 1971). In 1972–73 some resistant infestations existing in nurseries were treated with fonofos (Dyfonate) (USDA 1974).

In April 1973 the National Plant Board reviewed the status of the entire WFB program. The result of this review resulted in a public hearing to decide whether the federal quarantine of the pest should be continued. Following the hearing held on September 11, 1973 it was decided to retain the WFB federal quarantine and to add portions of the states of Kentucky, Missouri, and Texas to the regulated areas (U.S. General Services Administration 1974). In 1973, 7,400 acres were treated with residual insecticides at a cost of $11 per acre. In 1974–75, about 3,000 acres were treated with chlordane to control WFB (USDA 1976).

On May 19, 1975, it was announced in the Federal Register that the WFB quarantine was terminated effective June 30, 1975 (U.S. General Services Administration 1975). The reasons given for this decision were that "the whitefringed beetle had spread to much of its ecological limits and will continue to slowly spread in the United States. Accordingly, it has been determined that the Whitefringed Beetle Quarantine and regulations are no longer effective and necessary to prevent the spread of the white-fringed beetle" (ibid.). Also, "changes in agricultural practices have, in recent years, reduced the economic importance of this pest and made it possible for individual farmers to adequately apply available controls" (USDA 1975).

In 1979, a new infestation of WFB was discovered in Prince George County, Maryland. This was the first recovery of WFB in this state since the federal quarantine had been removed. The infestation was treated with chlorpyrifos (Dursban) at 0.82 gallons per acre once in the fall of 1979 and once during the summer of 1980, but recoveries of WFB continued. In July 1985, a second infestation was discovered in St. Mary's County, Maryland. No control activity against this infestation was performed (Charles Staines, Maryland Dept. of Agriculture, personal communication, 1987).

Evaluation

Since the early 1970s, the damage caused by WFB has been sporadic and localized. In the 1972–73 progress report of the Plant Protection and Quarantine Programs, damage from the pest was summarized as follows: "Patterns of damage to crops in infested areas show no distinct change. Damage, involving plant destruction and reduced yields in row crops, was reported as sporadic and occurring on small, scattered acreages in infested areas" (USDA 1974, 23). Herbert Womack of the University of Georgia, College of Agriculture, reported that damage occasionally will take place in watermelons and tobacco but less so in peanuts. He also stated that in 1984 in a localized area of Georgia (Miller County) significant damage occurred in 80 percent of the crops grown. Damage was not restricted to one or two crops but occurred in peanuts, corn, sorghum, soybeans, and various truck crops. When control is necessary, fonofos (Dyfonate) is recommended for use (H. Womack, personal communication, April 1985).

Richard Sprenkel at the University of Florida, reported that WFB is a sporadic pest of tobacco and peanuts in northern Florida, but that such grains as corn and sorghum seldom receive significant damage. If damage does occur it is usually restricted to a portion of a field (such as one corner) and rarely will the whole field be affected. In tobacco, damage is particularly critical during the establishment of new plants, but as the plants increase in size and vigor additional damage is less likely. Beginning in 1985, chlorpyrifos (Lorsban) applied at a concentration of 2–3 pounds active ingredient per acre has been registered for WFB control in tobacco (R. K. Sprenkel, personal communication, April 1985).

It has been ten years since chlordane and other chlorinated hydrocarbons have been applied for WFB control. Since the usual effectiveness of these applications is approximately five years, a resurgence of WFB damage may not be unexpected. The number of reports of WFB damage has increased in Georgia (David B. Adams, University of Georgia, personal communication, November 1988) and Florida (R. K. Sprenkel, personal communication, November 1988) over the last seven years, but it is too early to tell if a sustained increase in WFB damage will occur in the future. Other than activities by staff researchers, there are currently no ongoing sponsored research projects on WFB in Georgia or in Florida.

Since WFB eventually spread to its ecological limits in eastern North America and eradication was not achieved, the Federal-State Cooperative Control Program is considered a failure (Kim 1983). However, one may argue that it did not fail but rather was successful in "buying time." It is possible that the spread of WFB was greatly retarded due to the quarantine efforts. Popham and Hall (1958) compared the quarantine and subsequent spread of WFB with the introduction and spread of the Mexican bean beetle, *Epilachna varivestis* Mulsant. Following its introduction in 1920, no barriers were enacted against the Mexican bean beetle, and in fifteen years it had disseminated to almost every state east of the Mississippi River.

In contrast, the WFB control project may have delayed the eventual spread of the beetle to its expected ecological range for approximately twenty-five years. But what was the cost of the time that was bought? Although no estimates of the costs of insecticides, application equipment, and labor are available, considering the long-term commitment of this project, these costs must have been enormous. In addition to the up-front costs, there were other ecological costs. The development of insecticidal resistance was associated with the resurgence and spread of WFB in the mid 1960s. While parthenogenic reproduction may have served to prevent the rapid spread of genetic material conveying resistance throughout the WFB populations, resistance was still found to occur in four states six years after its initial discovery in Semmes, Alabama (Harlan et al. 1972).

In addition to the development of insecticidal resistance, the likely loss of fish and wildlife due to the mass broadcasting of DDT and dieldrin represented an additional environmental cost. Heavy losses of fish and wildlife caused by the use of dieldrin applied at a rate of 2 pounds active ingredient per acre were documented during the Fire Ant Eradication Program (Brown 1961). Dieldrin was applied at a rate of 3 pounds active ingredient per acre to suppress WFB populations. By 1977, all of the chlorinated hydrocarbons used against the pest were banned due to environmental hazards.

In the sense of delaying spread of the pest, the WFB control project may have been successful, but given the duration of the project, the commitment in materials and labor, and the environmental effects, the total cost was certainly very high.

Literature Cited

Boutwell, J. L., and D. L. Watson. 1978. Estimating and evaluating economic losses by white-fringed beetles on peanuts. *Bull. Entomol. Soc. Amer.* 24:157–59.

Brown, W. L. 1961. Mass insect control programs: Four case histories. *Psyche* 68:75–111.

Buchanan, L. L. 1939. The species of *Pantomorus* of American north of Mexico. USDA *Misc. Publ.* 341:1–39.

———. 1942. Four new species of white-fringed beetles (subgenus *Graphognathus*) from the southeastern part of the United States (Coleoptera: Curculionidae). *Bull. Brooklyn Entomol. Soc.* 37:107–10.

———. 1947. A correction and two new races in *Graphognathus* (white-fringed beetles) (Coleoptera: Curculionidae). *J. Wash. Acad. Sci.* 37(1):19–22.

Commonwealth Institute of Entomology. 1964. *Distribution maps of pests.* Map no. 179. London: Commonwealth Agricultural Bureaux.

East, R. 1980. Sampling white-fringed weevil (*Graphognathus leucoloma*) populations. *New Zealand J. Agric. Res.* 23:581–88.

Harlan, D. P., F. J. Bartlett, E. F. Pittman, G. R. Padgett, J. A. Mitchell, and Z. A. Shaw. 1972. Resistance to dieldrin in larval white-fringed beetles. *J. Econ. Entomol.* 65:1260–63.

Henderson, C. F., and L. J. Padget. 1949. White-fringed beetles distribution, survey, and control. USDA *bureau of entomology and plant quarantine,* E–779.

Kim, K. C. 1983. How to detect and combat exotic pests. In *Exotic plants pests and North American Agriculture,* ed. C. L. Wilson and C. L. Graham, 261–319. New York: Academic Press.

McClurkin, J. I. 1953. Methyl bromide fumigation of soil for destruction of white-fringed beetle larvae. *J. Econ. Entomol.* 46:940–44.

Metcalf, C. L., W. P. Flint, and R. L. Metcalf. 1962. *Destructive and useful insects.* 4th ed. New York: McGraw-Hill.

Minor, W. A., G. H. Collingwood, R. V. Heinkel, L. S. Hitchner, G. D. Humphreys, and R. E. Yung. 1952. *A report on the insect and plant disease control programs of the bureau of entomology and plant quarantine.* Washington, D.C.: USDA-ARS.

Ottens, R. J., and J. W. Todd. 1980. Leaf area consumption of cotton, peanuts, and soybeans by adult *Graphognathus peregrinus* and *G. leucoloma. J. Econ. Entomol.* 73:55–57.

Popham, W. L., and D. G. Hall. 1958. Insect eradication programs. *Ann. Rev. Ent.* 3:335–54.

Rohwer, G. 1953. White-fringed beetle surveys in noninfested states. *J. Econ. Entomol.* 46:1094–95.

Sailer, R. I. 1983. History of insect introductions. *Exotic plant pests and North Ameri-*

can agriculture, ed. C. L. Wilson and C. L. Graham, 15–38. New York: Academic Press.

Smith, J. L. 1948. Farming in Florida to beat the white-fringed beetle. *Agricultural extension service pamphlet.* Gainesville: University of Florida.

U. S. Department of Agriculture (USDA). 1938. Service and regulatory announcements *Bureau of entomology and plant quarantine,* 137:132–38. Washington, D.C.: U.S. Government Printing Office.

———. 1943. *Report of the administrator of agricultural research.* Washington, D.C.: U.S. Agricultural Research Administration.

———. 1947. *Report of the administrator of agricultural research.* Washington, D.C.: U.S. Agricultural Research Administration.

———. 1951. *Report of the chief of the bureau of entomology and plant quarantine.* Washington, D.C.: U.S. Government Printing Office.

———. 1953a. *Report of the chief of the bureau of entomology and plant quarantine.* Washington, D.C.: Government Printing Office.

———. 1953b. *Report of the administrator of agricultural research.* Washington, D.C.: U.S. Agricultural Research Administration.

———. 1958. *Cooperative economic insect report* 8:965. Washington, D.C.: Plant Pest Control Division, USDA-ARS.

———. 1959a. *Cooperative economic insect report* 9:570. Washington, D.C.: Plant Pest Control Division, USDA-ARS.

———. 1959b. *Cooperative economic insect report* 9:1017. Washington, D.C.: Plant Pest Control Division, USDA-ARS.

———. 1961. *Cooperative economic insect report* 11:137–38. Washington, D.C.: Plant Pest Control Division, USDA-ARS.

———. 1962a. *The white-fringed beetle: How to control it.* Leaflet 401. Washington, D.C.: Government Printing Office.

———. 1962b. *Cooperative economic insect report* 12:151–52. Washington, D.C.: Plant Pest Control Division, USDA-ARS.

———. 1963a. *Cooperative economic insect report* 13:16. Washington, D.C.: Plant Pest Control Division, USDA-ARS.

———. 1963b. *Cooperative economic insect report* 13:416. Washington, D.C.: Plant Pest Control Division, USDA-ARS.

———. 1963c. *Cooperative economic insect report* 13:565. Washington, D.C.: Plant Pest Control Division, USDA-ARS.

———. 1964. *Cooperative economic insect report* 14:390–91. Washington, D.C.: Plant Pest Control Division, USDA-ARS.

———. 1970. *Progress report.* Washington, D.C.: Plant Protection Division, USDA-ARS.

———. 1971. *Progress report.* Washington, D.C.: Plant Protection and Quarantine Programs, USDA-APHIS.

———. 1972. *Controlling White-fringed beetles.* Leaflet 550. Washington, D.C.: Government Printing Office.

———. 1974. *1972–73 progress report.* Washington, D.C.: Plant Protection and Quarantine Programs, USDA-APHIS.

———. 1975. USDA ends whitefringed beetle quarantine (press release). *United States Department of Agriculture News,* no. 1498-75, 16 May.

————. 1976. *Fiscal Year 1975 progress report*. Washington, D.C.: Plant Protection and Quarantine Programs, USDA-APHIS.

U.S. General Services Administration. National Archives and Records Service, Office of the Federal Register. 1974. Domestic Quarantine Notices, Subpart—Whitefringed Beetle, Quarantine and Regulations. *Federal Register* 39 (118): 21037.

U.S. General Services Administration. National Archives and Records Service, Office of the Federal Register. 1975. Domestic Quarantine Notices, Subpart—Whitefringed Beetle, Termination of Quarantine and Regulations. *Federal Register* 40 (97): 21693.

Warner, R. E. 1975. New synonyms, key, and distribution of *Graphognathus*, whitefringed beetles, in North America. USDA *Coop. Econ. Insect Report* 25:855–60.

Young, H. C. 1944. DDT against the white-fringed beetle and the velvetbean caterpillar. *J. Econ. Entomol.* 46:1094–95.

————. 1960. Methods for controlling white-fringed beetles. *Agric. Chem.* 15(8):36–37, 93.

Young, H. C., B. A. App, J. B. Gill, and H. S. Hollingsworth. 1950. White-fringed beetles and how to combat them. USDA *Circular 850*.

Young, H. C., B. A. App, and G. D. Green. 1938. The white-fringed beetle, *Naupactus leucoloma* Boh. USDA *bureau of entomology and plant quarantine*, E–420.

Young, H. C., and J. B. Gill. 1948. Soil treatments with DDT to control the white-fringed beetle. USDA *bureau of entomology and plant quarantine*, E–750.

9

Dutch Elm Disease

Donald R. Owen and Jill W. Lownsbery

Dutch elm disease (DED) is caused by the fungus *Ceratocystis ulmi* (Buisman) C. Moreau, which can be transmitted by many different species of bark- and wood-inhabiting beetles. Beetle species of the family Scolytidae are the most effective vectors (Lanier and Peacock 1981). All species of elm (*Ulmus*) and some species of closely related tree genera can be infected by the fungus. The precise origin of the pathogen is unknown, although it is believed to have originated in Asia. The disease was first noticed when it started killing native elms in Europe just after World War I. From Europe it was introduced to the United States, where it was first diagnosed in 1930 (May 1930). *Scolytus multistriatus* (Marsham), the principal vector of *C. ulmi* in the United States, was also introduced from Europe, apparently many years before the fungus. Today the disease is found throughout most of the United States and southern Canada. The greatest damage caused by the disease has been to the native American elm (*Ulmus americana* L.), which is planted widely as a shade tree throughout the east and midwest. By 1976 approximately 56 percent of the American elms present in urban areas had been killed by the disease (USDA 1977). There have been no successful programs to eradicate the disease.

Life History and Habits

The causal agent of DED, *Ceratocystic ulmi,* is an Ascomycete. Spore-producing structures of the fungus may be found in diseased trees, especially in the galleries of bark-inhabiting insects (fig. 9.1). The fungus grows in the xylem tissue of its host, particularly the vessels, resulting in a wilt disease

We thank B. W. Hagen, A. Haley, and A. H. McCain for reviewing this chapter and providing discussions on Dutch elm disease (USDA). We thank R. J. Campana and D. L. Wood for reviewing the chapter. We also thank F. W. Cobb, Jr., C. S. Koehler, J. R. Parmeter, Jr., and P. Svirha for discussions on DED.

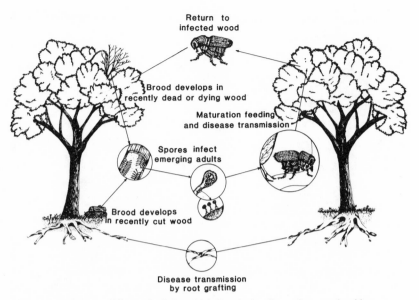

Return to
infected wood

Brood develops in
recently dead or dying wood

Maturation feeding
and disease transmission

Spores infect
emerging adults

Brood develops
in recently cut wood

Disease transmission
by root grafting

Figure 9.1. Diagram of the methods by which Dutch elm disease is transmitted between host trees (adapted from University of California, Division of Agricultural Sciences, Leaflet X122, undated).

that is characterized by (1) obstruction of infected vessels through development of tyloses, (2) cessation of water flow to leaves, (3) death of leaves and succulent shoots, (4) spread of the fungus downward into the large stems and roots, and (5) death of the tree. The speed with which the disease overtakes its host varies depending upon the elm species, tree nutrition, soil moisture, temperature, tree vigor, and the mode and site of inoculation (Van Alfen and MacHardy 1978). In some cases infections may be limited to single branches. In addition to being a virulent pathogen, the fungus may also grow as a saprophyte in the dying wood of trees infested with bark beetles (Rankin, Park, and Collins 1941).

The fungus can move from one elm to another either by growing down into the root system of its diseased host and across a root graft to a healthy tree or via transmission by a bark beetle (fig. 9.1). In some areas as much as 50 percent of the disease spread is via root grafts (Cuthbert, Cannon, and Peacock 1975). In North America two species of bark beetles (Coleoptera: Scolytidae) are important vectors of *Ceratocystis ulmi*.

Scolytus multistriatus, the smaller European elm bark beetle, is the most important vector in the United States, where it was first reported in 1909 after its introduction from Europe (Chapman 1910). Since that time its range has expanded to cover all areas where elms are native or planted (fig. 9.2). It breeds in the bark of dead or dying *Ulmus* spp. or *Zelkova* sp. Adults

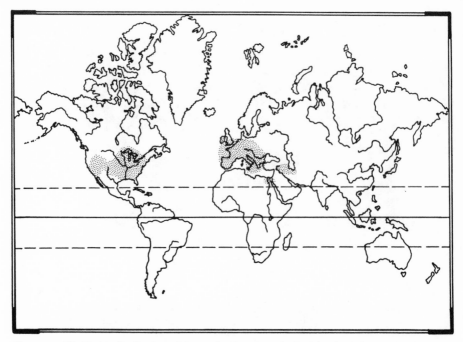

Figure 9.2. Worldwide distribution of the smaller European elm bark beetle, *Scolytus multistriatus* (adapted from Commonwealth Institute of Entomology 1975).

construct characteristic egg galleries in the phloem-cambial tissues and engrave the xylem somewhat. Larvae also feed in the phloem-cambial region, making individual mines perpendicular to the parental galleries. The number of generations produced per year varies from two in colder northern regions to three or a partial third in such warmer regions as California or southern New Jersey.

Hylurgopinus rufipes Eichh., the native elm bark beetle, occurs throughout the range of elm in the East and is the most important vector in parts of Canada ("Review of Dutch elm disease" 1964). It breeds in the bark of dead or dying elm trees, producing one and a half generations per year. Egg and larval galleries are mined in the phloem-cambial tissues of the host.

Both the asexual and sexual spores of the fungus are borne in a sticky matrix that adheres to the bodies of beetle vectors. Newly emerged beetles can transmit the fungus when they feed on an uninfected host. In the case of the European elm bark beetle, infection usually occurs in spring when newly emerged adults feed in the twig crotches of healthy elms before flying to a suitable host for breeding. Similarly, transmission by the native elm bark beetle usually occurs when adults crawl or fly from their overwintering sites in the lower boles of elm trees to small limbs in the crown where they feed on the phloem and score the xylem (Thompson and Matthysse 1972). Some transmission also occurs in the overwintering sites of these beetles.

Two conditions must be met before the fungus spores can infect. The feeding wound must reach the xylem, and the microclimate at the wound site must be humid ("Review of Dutch elm disease" 1964). The time of year that the feeding occurs is also important in disease development, since the fungus travels most rapidly in the large diameter vessels that are produced in the spring.

For additional information one may consult the reviews given in Sinclair and Campana (1978) and Stipes and Campana (1981). A bibliography is provided by Laut and Shomaker (1979), with addendum by Laut and Stieger (1980).

Epidemiology and Ecology

The pattern of disease spread within a particular disease center has been summarized by Sinclair (1978b). Trees closest to the sites of beetle breeding are fed upon most often and become infected first. In the first two to four years very few trees become infected because the number of beetles contaminated with the fungus is relatively low at that time. Later the infection rate accelerates to an annual level of between 10 and 20 percent of the original elm population. After ten to fifteen years most elms in the area are killed. The rate of spread in northern areas is slower.

A number of factors affect both the development of new disease centers and the spread of existing ones (ibid). Probably the most important factor is the amount of dead or dying elm material for the bark beetle vectors to utilize. Climate and latitude have been found to affect fungus development, beetle development, and the synchronization between the peak emergence of elm beetles and the time of peak susceptibility of the elms in the spring. The most favorable climate for spread of the disease appears to occur in the midwestern United States. Though elms are abundant, spread of the disease in the northern United States and southern Canada has been much slower.

The proportion of the two vector species present at a particular locale also appears to be important. Comparisons of infection rates where both species occur together versus areas where only *H. rufipes* occurs suggest that *S. multistriatus* is more likely to score the xylem while feeding during the period of peak elm susceptibility than is *H. rufipes*.

The amount of root grafting between adjacent elms is important, particularly in areas where beetle suppression programs occur (Cuthbert, Cannon, and Peacock 1975). In such situations as much as 50 percent of the disease spread may be attributable to root graft transmission. Generally elms of the same species planted ten meters or fewer apart are likely to possess root grafts (Himelick and Neely 1962).

The presence of fresh wounds, such as those created by pruning, has been shown to attract bark beetles (Hart et al. 1967). The beetles sometimes tunnel into the phloem at the edges of the wounds and, if contaminated with *C. ulmi* spores, can transmit the fungus.

A number of natural enemies attack the two bark beetles species; these include various insect predators and parasites, nematodes, birds, and fungi (Peacock 1973). Wherever the disease is epidemic, however, these agents do not affect enough beetles to influence the infection rate (Sinclair 1978b).

The relative proportions of the different elm species present can be important. Different elm species vary greatly in their susceptibility to the disease. The American elm is the most susceptible species, whereas other elms, particularly the Asiatic elms, are more resistant (Sinclair 1978a). On the other hand, the pathogenicity (the relative capability of a pathogen to cause disease) of the fungus also varies. Certain strains, termed aggressive, can cause wilt and dieback in normally resistant elms (Brasier and Gibbs 1973). Nonaggressive strains cause conspicuous symptoms only in susceptible elm species, like the American elm. In addition, there are other strains displaying intermediate pathogenicities. All three types may be found in North America (Shreiber and Townsend 1976).

Damage and Control

Losses of elms planted as ornamental and shade trees in the United States have been dramatic. An estimated 40 million elms have been killed since 1930. Annually, it is not uncommon for 10 percent or more of planted elms to be killed by DED in areas where the disease is well established and control is not practiced. On the average, losses of greater than 75 percent of the original elm population have occurred in areas where the disease has been present the longest. Costs to control the disease (aside from research) have been substantial. In 1977 municipal control programs were estimated to cost a total of $30 million per year (USDA 1977). As of 1984 the disease had been discovered in all but six (Florida, New Mexico, Arizona, Nevada, Hawaii, and Alaska) of the fifty states (R. J. Campana, personal communication, 1984).

Two factors that have contributed to the substantial losses observed in the United States are the large number of highly susceptible hosts and the climatic suitability for effective vector transmission of the disease-causing fungus. American elms in the eastern and midwestern United States, where this host is both native and widely planted, have suffered the highest levels of mortality. There are no elms native to the western United States

(Sargent 1961), but numerous species have been planted there. Although all elms are susceptible to DED, the least susceptible or most resistant are the Asian species. It is not unusual for these species to be infected without dying and to show negligible symptom development (Sinclair 1978a).

In Europe, losses due to DED have also been extensive. First reports of the disease in mainland Europe were in 1918 and 1919. In 1927 the disease was found in England (Gibbs 1981). Probably the greatest threat to Europe has been the introduction of an aggressive strain of the fungus, which is believed to have been introduced from Canada sometime during the 1960s. An epidemic attributed to this strain occurred in southern England, killing an estimated 70 percent of the elm population from 1971 to 1978 (ibid).

A considerable amount of research has been conducted on the control of DED, but no cure for the disease has been found. Specific control methods may be aimed at either the fungus causing the disease, the beetles that transmit it, or both. Some commonly used methods include detection and sanitation, application of dormant insecticidal sprays, and severing of root grafts.

The key to any successful control effort is finding the fungus and its vectors by conducting a detection survey. This is done by searching for all elm material that is likely to be infected by *C. ulmi* or that may serve as a breeding site for vectors, including disease trees, weakened trees, and cut elm wood (Collins, Parker, and Dietrich 1940). Once located, this material is promptly disposed of by burning or burying, thus destroying sources of the disease and preventing its spread. The effectiveness of survey and sanitation depends on how thoroughly and frequently the programs are conducted. Frequency is especially important because beetle infested material must be destroyed before the insects have the chance to emerge and carry fungal spores to new hosts. Conducting surveys three times per year and following with prompt removal of diseased trees has been shown to significantly reduce losses and tree removal costs when compared to programs where survey and tree removal (several months or more later) are conducted only once per year (Barger 1977; Cannon, Barger, and Worley 1977).

Although detection and sanitation may greatly reduce losses caused by the disease, they do not completely eliminate disease spread. Noninfected trees may be protected by spraying them with insecticide, and if a nearby tree is infected, by severing root grafts between the trees. Insecticide treatments are designed to prevent beetles from inoculating and breeding in trees. To be most effective, insecticides must be applied while the insects are dormant—that is, before the spring flight—and must remain operative throughout the flight period or be periodically reapplied. Because it is not usually practical or possible to remove below-ground portions of diseased trees, it may be necessary to sever root grafts between diseased and healthy trees in order to prevent disease transmission. Grafts can be broken mechanically by digging a trench between the trees or chemically by localized application of a soil fumigant that kills a portion of the roots.

Unfortunately, American elm does not cross with other elm species, and hence attempts to cross it with more resistant species have been unsuccessful. Most success has been in crossing species other than American elm with Asian elm species. A drawback of these crosses is that they are not ornamental substitutes for the American elm (Lester 1978). Attempts to develop resistant strains of American elm have shown promise but have not yet been proven successful. Even though resistant trees have been found, their progeny have not necessarily been resistant (Sinclair et al. 1974). Two factors that make it difficult to determine resistance are that (1) progeny of resistant parents are less likely to exhibit resistance if inoculated at a young age, and (2) if a tree is inoculated frequently enough, its resistance may be overcome (Sinclair and Larsen 1980). American Liberty elms are selected clones of American elm that possess resistance as mature trees, and show promise as replacements for American elms lost to DED (Eugene B. Smalley, personal communication, 1984).

When the costs of using no control versus control were compared for midwestern and eastern U.S. municipalities over a fifteen-year period, it was found that costs with control were much less than costs without. Savings ranged from 37 to 76 percent (Cannon and Worley 1980). This difference was due to the large costs of tree disposal and property loss under the no-control programs. Without control, 85 to 90 percent of the elms could be expected to die within twelve years. Thus, when a high level of tree mortality due to DED is expected, control programs are justified from an economic standpoint.

History of Eradication Efforts

Dutch elm disease was first reported in the United States in 1930 from an American elm in Cleveland, Ohio (May 1930). In 1933 a much larger infestation of the disease was discovered in New Jersey, prompting the establishment of a formal eradication program that remained in effect until 1941, when funding was discontinued as the country prepared to enter World War II. Following the war, most control efforts were conducted at the community level and were concerned largely with control and sanitation.

After the discovery of DED in Ohio, the Ohio Agricultural Experiment Station, in cooperation with the Department of Agriculture, instituted an immediate search for other infected trees (Beattie 1932). Surveys were also carried out in Indiana, Illinois, Missouri, Kentucky, and West Virginia, but no diseased trees were found in these states. During 1930 and 1931, the greater part of Ohio was searched. Three infected trees were found in

Cleveland and one in Cincinnati, Ohio (Beattie 1934). The following year four more diseased trees were located in Cleveland. All infected trees discovered were destroyed (Hepting 1977). The survey in 1932 did not locate additional trees, and it was concluded that the disease had been eradicated from Ohio (Beattie 1934).

The source of the disease was unknown at that time, though Beattie (1932) speculated that the disease may have been brought in on box lumber or elm leaves shipped from an infected region. Since movement of elm nursery stock from Europe to American had been prohibited by a quarantine on June 1, 1919, infection from a European source seemed unlikely.

In 1933 more diseased trees were found, and the probable source of the introduction was located. During the summer, a park foreman found a dying elm in Maplewood, New Jersey (Hepting 1977). In July of the same year an inspector of the Bureau of Plant Quarantine at the port of Baltimore, Maryland, discovered elm burl logs that were infested with *Scolytus* beetles (Beattie 1934). The logs were imported from Europe and were intended for production of veneer. Further examination disclosed the presence of *C. ulmi* in the wood. Subsequently, similar shipments were intercepted at New York, Norfolk, Virginia, and New Orleans, Louisiana. Again *Scolytus* beetles were found, including both *Scolytus scolytus* F. and *S. multistriatus*. *C. ulmi* was isolated from three of the shipments. Federal forest pathologists reviewed the combined import records of the Departments of the Treasury and Commerce, checking for any shipments that might have included elm. They found that European elm burl logs had been shipped to various places in eleven states (Hepting 1977). These shipments started as early as 1925 and continued through 1934, with eleven shipments entering at New York City, twenty-two at Baltimore, nine at Norfolk, and sixteen at New Orleans (Liming 1937). Later it was noted that the occurrence of the disease in nearly all cases was traceable to the movement of these imported elm logs (Rankin 1937). A quarantine was subsequently issued and went into effect on October 21, 1933, regulating the entry of such logs into the United States (Beattie 1934). It also forbid or regulated the importation of other elm or related plant parts into the United States.

Soon after the confirmation of the disease in New Jersey, it was also discovered in New York, Connecticut, and Maryland (ibid). The Bureau of Entomology and Plant Quarantine moved to organize a campaign of control and eradication. The objectives of this program were first to control the disease where it occurred in epidemic form, such as around New York City, and then to eliminate it in outlying areas. Cooperation with the state departments of agriculture in New York, New Jersey, and Connecticut was provided through special state appropriations and enactment of state plant pest control laws (American Forestry Association 1938). In 1934 the area of infestation was defined as those lands within a fifty-mile radius of Columbus

Circle, including most of northern New Jersey, Rockland and Westchester counties in New York, and Fairfield county in Connecticut (White 1935). Outside of this area a protective zone approximately ten miles across was established (Brewer 1941). The plan was to systematically inspect every elm tree in the infested area at monthly intervals and to check the border zone at least once each year. In addition, all elms in both the infested area and border zone that were more than one-half dead were to be tagged for winter removal (Rankin 1937). Certain outside areas, principally port and mid-western cities that handled imported logs and about twelve thousand miles of railroad involved in hauling the logs, were also to be checked (Liming 1937). Labor was provided through relief agencies such as the Civilian Conservation Corps and the Works Progress Administration (Hepting 1977). In 1934 the Dutch Elm Disease Lab was moved from Wooster, Ohio, to Morristown, New Jersey, where it conducted research basic to the erad-ication effort.

In 1933, 820 diseased elms were located and removed following surveys concentrated in urban areas. Undeveloped areas and the ten-mile protection zone were not checked with any degree of thoroughness (American For-estry Association 1938; Liming 1937). The removal of devitalized elms began in the winter of 1933 and continued in 1934. Unfortunately, tree removal occurred too late to prevent the emergence of disease-carrying beetles. In 1934, 6,867 diseased trees were located and destroyed. This work was fi-nanced mostly through emergency appropriations from the federal govern-ment totaling $351,020 and in part from appropriations from the state of New Jersey amounting to $5,000 (American Forestry Association 1938). As in the previous year, scouting was concentrated in urban areas, and diseased trees were not removed before beetle emergence.

Funding and scouting efforts increased in 1935 and 1936, with $786,186 spent in 1935 and $5,554,066 in 1936 (ibid). For the first time extensive scouting was undertaken in undeveloped areas and in the protective zone (Liming 1937). Further, "worthless" woodland elms were removed begin-ning in fall 1936. That effort became feasible with the development of a new technique costing only one-tenth as much as the older method. Using this technique, called silvicide, trees were girdled with an axe and poison applied directly to the sapwood, which was then covered with a heavy cloth ban-dage. The loose bark was than tacked down over the cloth to provide protection against the weather (ibid).

The situation remained about the same in 1937, but it worsened in 1938. A total of 18,152 diseased elms were located, mainly in New Jersey (Brewer 1941). Many people, including some scientists as well as members of the National Conference on Dutch Elm Disease Eradication, believed the battle had been lost. Their criticisms of the federal project included all phases of the eradication program. Subsequently, in 1939 an increase in funding made more extensive surveys possible (ibid). The border zone was increased from ten to fifteen miles and checked rather intensively. An advance zone extending about a hundred miles beyond the outer edge of the protection zone was checked less intensively, and the area located from a hundred to a hundred and fifty miles beyond was scouted lightly. Following that effort, the known infested area was increased by over twenty-three hundred square miles (ibid). A similar program was planned for 1940; however, the amount of money appropriated was reduced and funding was late, so the program was scaled down. The main area of effort was the fifteen-mile-deep protection zone, which at that time covered fifty-five hundred square miles. Only about 4000 diseased trees were found (ibid). In 1941 the program was discontinued because of the war. After World War II the effort was revived, but funding was poor and not well directed; thus, control was switched to the community level (Wilson 1975).

From 1933 to 1941, nearly $27 million were spent (Hepting 1977) and 2.6 million elms were sampled, with 60,975 cases confirmed (Brewer 1941; Hepting 1977). Sanitation operations removed 4.25 million elms, but the area of infestation had increased from 2,464 square miles in 1934 to more than 10,900 square miles in 1940 (Brewer 1941). The project was plagued with difficulties from the start.

Many of the problems encountered during the project were related to the Works Progress Administration (WPA). Nearly 90 percent of all expenditures came from this agency (Wharton 1938), and its regulations stated that the labor provided through its moneys must come from relief rolls. Unfortunately, these workers were unskilled, and sometimes unsuited for the assigned tasks (Brewer 1941). In addition, there was an eighteen-month limit on their length of employment, which probably lowered worker morale (ibid). Another problem was that the highest concentration of relief workers occurred in heavily populated areas. Workers could not be employed except near their registration points. Thus, there were difficulties in placing workers in the appropriate areas (Wharton 1938). This led to the suggestion that unless workers were obtained from other sources the program would fail (ibid).

The timing and amount of money allocated through these relief agencies was also a problem. Often the funding came too late to provide for adequate early season scouting. Funds should have been available by May 1, so that the crews could be trained and ready to go between May 30 and June 15 (Brewer 1941). Frequently this was not the case. In 1940, the agricultural bill authorizing the allocation of moneys was passed in late June. Thus, it was July before the WPA allotments became available (ibid). The amount of money each year was also uncertain. For example, in the last part of June 1938, all WPA workers had to be laid off because of uncertain funding for the new fiscal year. Then, even though the funds were made available on July 1, it was not known until August how they would be expended (Wharton 1938).

There were a number of other notable problems. First, early surveys concentrated in the urban areas, while the outlying areas and the protection zone were surveyed superficially. As a result, the disease was left unchecked and possibly even undetected in the rural areas of the main infestation and protection zones. Wharton (1938) suggested that this may have defeated eradication efforts. The second problem the program did not address was the breeding of beetles in downed elm wood, where the fungus could occur as a saprophyte. Efforts were directed only at removing standing diseased or devitalized elms and preventing the importation of cut elm wood. A third problem was the spread of the fungus by root graft because only the above-ground portions of diseased trees were removed and destroyed. Researchers were concerned about the length of time the fungus could survive in the root system of a cut tree, but the funds for experimentation were limited. Last, there were equipment problems, especially at the start. Liming (1937) complained that the low quality of the cutting tools in combination with the inadequate number of vehicles made work crews less efficient.

Even if all of the above problems did not occur, it still would have been difficult to contain or eliminate the fungus. The existence of between five and ten million American elms in New Jersey, New York, and Connecticut made it unlikely that all diseased trees could be located. The distribution of the disease was more widespread than initial surveys had indicated.

The attempt of the state of California to eradicate DED is the most recent example of a major coordinated effort to eliminate this disease. Although DED had never been eradicated from a large geographical area, incentives that made the attempt seem worthwhile were present in California. Presently, DED is well established in the San Francisco Bay Area, and although efforts to eradicate the disease have been abandoned, control measures are still in effect.

In 1947 the federal government revoked its domestic quarantine regulating the transportation of elm material within the country because the quarantine was considered impractical. That same year California enacted its Exterior Quarantine 21, which prohibits the importation of *Ulmus, Zelkova,* and *Planera* spp. plants and plant parts (except seeds and lumber without bark) unless they are certified to have originated in an area free of DED (Ryan et al. 1969). Despite this quarantine, the first discovery of *Scolytus multistriatus* in California was made in 1951 in San Jose, and a year later it was discovered in southern California. There was no evidence, however, that *Ceratocystis ulmi* had been introduced along with the beetle (Brown and Eads 1966).

Although the principal vector of DED became well established in the state as early as the 1950s, major concern over the possibility of DED being introduced did not develop until the 1970s, when DED was detected closer to California. Partridge and Weir (1974) reported finding DED in Oregon in 1973, which had apparently spread there from Idaho. The infections in Oregon were approximately 240 air miles from the northern border of California (DED Task Force 1974). As the disease moved westward, the threat of a DED introduction into California increased. In response to this threat, a state DED task force was established, and a detailed action plan prepared (ibid), which included the following warning: "The spread of the disease among the state's estimated two million elm and *Zelkova* trees will be rapid and devastating unless control efforts are initiated immediately. To wait until the disease appears in California before taking action will mean that the battle has been lost before it has begun." Present in the plan were recommendations relating to exclusion of the disease from the state, detection of initial infections, vector suppression, public relations, and prompt containment and eradication of DED should it be discovered. Above all, the plan emphasized the immediacy and potentially destructive nature of the DED threat.

Ironically, DED was discovered in California only one year later, before funding for and implementation of the task force's action plan could be made. The disease was initially found in three counties (Sonoma, Napa, and Santa Clara) during July–September 1975. These counties or parts thereof were quickly put under quarantine. On September 24, 1975, Assembly Bill 1874 was made law, authorizing the California Department of Food and Agriculture (CDFA) to begin an eradication effort. Since then, about $1.4 million has been appropriated annually for the eradication and control of DED in California (California Department of Forestry 1982).

Despite the firm commitment of CDFA to an eradication approach, the rationale behind such an approach was seriously questioned by some, particularly scientists at the University of California (Fields W. Cobb and John R. Parmeter, personal communication, 1983). Foremost among the criticisms was that no proven eradication system existed. There was no example anywhere of a deliberate, systematic effort that had succeeded. Additionally, the disease was not seen as much of a threat to the elm population of California for two reasons. First, there are no native elms present in the state, and only about 8 percent of the estimated 1.5 to 2 million planted elms in California are American or European elms. The other 92 percent are Asiatic species (University of California Cooperative Extension 1976), which are much less susceptible to the disease. Regional estimates may, however, differ substantially from this statewide estimate. For example, in Napa and Marin counties the fraction of American and European elms is approximately 50 percent (Bruce W. Hagen, personal communication, 1983, California Department of Forestry, Dutch Elm Disease Project). Second, The Mediterranean climate of California is probably not very conducive to disease spread because warm, dry conditions prevail throughout most of the state for much of the year. Kais, Smalley, and Riker (1962) demonstrated that low relative humidity at the wound site greatly inhibits the infection process and that trees existing under dryer conditions are less susceptible than trees with a more adequate water supply. For these reasons, damage from DED in California might not become extensive. Even if DED could be eradicated from the state, was the cost of such a project justified in terms of probable losses, and might not the disease be reintroduced repeatedly in the future?

Partly as a response to this controversy, a joint committee of University of California (UC) scientists and CDFA personnel was established in 1976 to confer on the progress and direction of the eradication program. All of the UC scientists chosen for the committee were initially in favor of eradication. The consensus of the committee was that it was better to make an attempt at eradication than never to know if it was possible. At the time, the disease was believed to be confined to a small geographic area and recently introduced into the state (Arthur H. McCain, personal communication, 1983).

Campana (1978) presented some pros and cons of attempting DED control in California, based on his observations of the disease in California and the eastern United States. He pointed out that the disease is likely to spread more slowly in California because of host and climatic factors, but that the disease will be more difficult to detect, partly because of the lack of symptom development in less susceptible species. Whereas slow disease spread would favor the eradication attempt, difficulty in detecting the disease would hinder it. Provided that the disease was discovered shortly after its introduction and before it had a chance to become well established, there seemed to be a chance that it could be eradicated. This is most likely why the DED Task Force (1974) emphasized the importance of survey and early detection: "Prompt detection of the first incipient outbreak of DED in California is considered vital to the success of our control program."

In 1975 when DED was first discovered in three Bay Area counties, an intensive program of survey and detection, tree removal, sanitation, vector suppression, and other control methods was initiated in quarantined areas. In that year, an estimated 884 diseased trees were removed. Each year after its initial discovery and until 1979, the disease was found in additional counties. By 1979, a total of 1800 diseased trees had been removed from eight Bay Area counties (California Department of Forestry 1982). As survey and detection activities improved, a truer picture of the distribution of the disease was revealed. It became evident that the disease was not spreading at a tremendous rate, but rather it was being discovered in areas where it had already been present for some time.

Jones, Krass, and Sava (1978) reported the isolation of *C. ulmi* from the 1963 and 1964 growth rings of an English elm from Kenwood, California, that had died from DED in 1976. The tree had reacted to the earlier infections and apparently "walled" them in, proving that the infections had indeed occurred in the same years as the growth rings indicated (or earlier). Campana, French, and Locatelli (1981) isolated *C. ulmi* from pre–1978 annual rings of over 100 elms in California, with the oldest isolation being a 1958 ring. Similarly, Tidwell and Sava (1982) isolated *C. ulmi* from the 1965 annual ring of two trees (Sonoma and San Mateo Counties) and the 1953 ring of a single tree (San Mateo County), as well as making positive isolations from pre–1975 rings of numerous trees. Based on the 1953 and 1958 annual ring-isolations, it is possible DED may have been introduced into California along with or soon after the first introduction of *S. multistriatus*. Tidwell and Sava (1982), however, expressed reservation over the significance of the 1953 growth-ring isolation because of possible movement of the fungus into this ring from younger rings. Regardless of exactly when DED was first introduced into California, it was apparently well established in the Bay Area by the time of its first discovery in 1975.

By the end of 1977, DED had been found in six Bay Area counties and in Los Angeles county, and UC members of the joint UC/CDFA committee on DED were expressing serious doubts about the likelihood of eradicating DED from California. Not only was the disease more widespread than earlier thought, but there was concern that more trees were being removed than might be necessary. That is, all trees with confirmed disease were being cut down and disposed of by CDFA, but evidence indicated that a certain proportion of these could recover. This particular situation is somewhat problematic, however, because DED-infected trees that are not disposed of may contribute to further spread of the disease, and it would be difficult to predict which trees would survive. In 1978, isolations from *S. multistriatus* revealed that viable spores of *C. ulmi* could be recovered from beetles trapped at 41 percent of the sites (areas within a 1,000-foot radius of a confirmed diseased tree that has been removed) at which no diseased elm trees or wood had been found for two years. Hence, inoculum was still being carried by beetles in quarantined areas. Based on this and other information given in a CDFA status report, the UC members of the joint DED committee concluded that DED could not be eradicated from California, a conclusion opposite than that presented in the CDFA report. In a 1979 internal report, the UC members of the joint DED committee unanimously recommended that the eradication effort be discontinued in favor of a control program.

In November 1981 the DED Project was transferred from the California Department of Food and Agriculture to the California Department of Forestry, a move that formally ended the eradication attempt. From an operational standpoint, however, the goal of eradication has been unrealistic since 1979.

As of 1982, DED has been discovered in nine Bay Area counties and remains limited to this area. Currently the project's primary objective is to keep losses in quarantined areas to less than 1 percent per year (California Department of Forestry 1982), an objective that is being successfully met (B. W. Hagen, personal communication, 1984). It is anticipated that, as the incidence of DED-infected trees declines, less money will be needed for survey and detection, the most expensive parts of the program's budget. In Los Angeles County, one case of DED was confirmed in 1977. Since then, DED has not been found in southern California (California Department of Forestry, unpublished report).

Evaluation

Eradication attempts have not been successful with DED anywhere. The California program was the most recent attempt and therefore will be examined in some detail. When DED was discovered in California there was no way of knowing how widespread it initially was and therefore how easily it could be contained and eliminated. Certainly if the disease had been newly established with a limited distribution then eradication was a possibility. But neither of these appeared to be the case.

By the end of 1978 the following evidence was sufficient to warrant abandoning the goal of eradication: (1) the disease had been discovered in seven Bay Area counties, indicating it was widespread; (2) *C. ulmi* had been isolated from the older growth rings of numerous trees and as far back as 1953, indicating the disease had been well established yet remained undetected; (3) The fungus could be isolated from beetles even in areas where the disease had been "eradicated," indicating that the causal agent was more prevalent than the presence of the disease suggested.

Aside from the fact that DED was well established in California before its discovery, eradication may have been impossible or impractical for additional reasons. Some questions remain unanswered: (1) Would it ever be possible to eliminate DED without also eliminating its principle vector, *S. multistriatus?* (2) Given California's climate and relatively small population of American and European elms, DED would not likely develop into a widespread economic problem. Therefore was an expensive eradication program justifiable? (3) Even if DED were eliminated from California, are not reintroductions of the disease a likely event?

It is unlikely that DED could have been eradicated without also eradicating its principal vector, and eradication of *S. multistriatus* would have been virtually impossible. The beetle provides the only effective means of long distance transmission of the causal agent in California, although people may transport wood infested with the beetles. Elm populations are well dispersed in California; thus, the beetle is essential to disease spread and maintenance. Isolations from *S. multistriatus* in areas where DED was known to exist yielded viable *C. ulmi* from only a small proportion (fewer than 3 percent) of the total number of beetles trapped during 1980 in California (Tidwell, Kiser, and Butler 1980). The significance of such an apparently low level of transmission is not easily interpreted. First, when vector populations are high, even a very small proportion of the total population will translate into a large number of individuals (for example, 3 percent of 1 million is 30,000). Second, more beetles will likely be carrying spores than isolations indicate. The likelihood of initial contamination of the beetle with *C. ulmi* spores, the survival of these spores on the beetle, and the effectiveness with which a beetle inoculates a host are all important yet largely unknown factors. Even though *S. multistriatus* appears to infrequently carry *C. ulmi* spores in California, the ability of this insect to perpetuate the disease should not be underestimated.

Much published information exists for DED in other parts of the United States and the world, but strict application of these findings to California conditions would be inappropriate. The host and climatic conditions in California are quite different from those areas where DED has been most damaging—that is, eastern and midwestern United States, England, western and northern Europe. Because DED was present in California up to twenty-two years before it was discovered, it is reasonable to assume that the disease is slower in its development and not capable of causing as much damage in California as in these other areas. In the 1974 CDFA action plan, a devastating scenario was assumed should DED be introduced. There was no mention that DED might not be a serious threat, even though this was a reasonable possibility. Nor was there an economic justification for the programs proposed in the plan. In these respects, the plan was deficient and biased.

Although DED does not appear to pose a threat to the majority of elms in California, certain communities within and around the San Francisco Bay Area are likely to benefit from DED detection and control measures. These would be communities containing a large number of European and/or American elms and therefore having considerable property of aesthetic value at stake. Because DED has remained limited to the Bay Area for such a long period of time, the ability of the disease to spread to and develop in warmer, drier parts of the state is questionable.

There is very little published information on DED in California. Because conditions in California are unique in many respects, research on DED could provide valuable answers to many questions. Foremost among these questions is why is *S. multistriatus* so widespread in California, while DED is not? Is *C. ulmi* actually much more widespread in California than just the Bay Area? Do host and other environmental conditions limit disease development or expression? What features of the fungal/vector relationship influ-

ence disease spread? Control efforts in California and elsewhere could benefit from research on these topics. Research programs could be designed to
help control personnel evaluate their progress and to suggest future courses
of action. Often we do not know why a particular treatment succeeded or
failed, or to what extent our actions are changing the natural course of
events.

Literature Cited

American Forestry Association. 1938. The American elm, its glorious past, its present
    dilemma, its hope for protection. Washington, D.C.: Amer. For. Assoc.
Barger, J. H. 1977. Improved sanitation practice for control of Dutch elm disease.
    USDA *For. Serv. Res. Pap.* NE–386.
Beattie, R. K. 1932. Dutch elm disease survey indicates this disease not widespread in
    this country. *U.S. Dept. Agric. Yearbook* 1932:178–81.
———. 1934. Dutch elm disease now serious around New York: Entered Country in
    Logs. *U.S. Dept. Agric. Yearbook* 1934:188–90.
Brasier, C. M., and J. N. Gibbs. 1973. Variation in *Ceratocystis ulmi*. *Proc. Interna-*
    tional Union of Forestry Research Organizations (IUFRO) Conf., Sept., 53–76.
Brewer, E. G. 1941. The fight for elms: Review of progress made in the campaign
    against the Dutch elm disease. *Amer. For.* 47:22–25.
Brown, L. R., and C. O. Eads. 1966. A technical study of insects affecting the elm tree
    in southern California. *Univ. Calif. Agr. Exp. Sta. Bul.* 821.
California Dept. of Forestry. 1982. *Dutch Elm Disease Project Handbook.*
Campana, R. J. 1978. Comparative aspects of Dutch elm disease in eastern North
    America and California. *Calif. Plant Path. U. C. Coop. Ext.* 41:1–4.
Campana, R. J., A. French, and R. Locatelli. 1981. Isolation of *Ceratocystis ulmi* in
    California from elms with buried infections from previous years. *Phytopath.*
    71:207.
Cannon, W. N., Jr., J. H. Barger, and D. P. Worley. 1977. Dutch elm disease control:
    The economics of intensive sanitation. USDA *For. Serv. Res. Pap.* NE–387.
Cannon, W. N., and D. P. Worley. 1980. Dutch elm disease control: Performance and
    costs. USDA *For. Serv. Res. Pap.* NE–457.
Chapman, J. W. 1910. The introduction of a European scolytid (the smaller elm bark
    beetle, *Scolytus multistriatus* Marsh.) into Massachusetts. *Psyche* 17:63–68.
Collins, D. L., K. G. Parker, and H. Dietrich. 1940. Uninfected elm wood as a source
    of the bark beetle (*Scolytus multistriatus* Marsh.) carrying the Dutch elm disease
    pathogen. *Cornell Univ. Agr. Exp. Sta. Bull.* 740.
Commonwealth Institute of Entomology. 1975. *Distribution maps of pests*. Map no.
    347.
Cuthbert, R. A., W. W. Cannon, Jr., and J. W. Peacock. 1975. Relative importance of
    root grafts and bark beetles to the spread of Dutch elm disease. USFS *Res. Note*
    NE–206.
Dutch Elm Disease (DED) Task Force (A. M. French, chairman). 1974. *An action plan*

against Dutch elm disease. Sacramento: Calif. Dept. of Food and Agric. Div. of Plant Ind.

Gibbs, J. N. 1981. History in Europe and Asia. In *Compendium of elm diseases,* ed. R. J. Stipes and R. J. Campana. St. Paul, Minn: Am. Phytopath. Soc.

Hart, J. H., W. E. Wallner, M. P. Caris, and G. K. Dennis. 1967. Increase in Dutch elm disease associated with summer trimming. *Plant Dis. Reptr.* 51:476–79.

Hepting, G. H. 1977. The threatened elms, a perspective on tree disease control. *J. of For. His.* 21(2):90–96.

Himelick, E. B., and D. Neely. 1962. Root grafting of city-planted American elms. *Plant Dis. Reptr.* 46:86–87.

Jones, R. K., C. J. Krass, and R. J. Sava. 1978. Isolation of *Ceratocystis ulmi* from 14-year-old annual rings of English elm in California. *Plant Dis. Reptr.* 61(11):994–95.

Kais, A. G., E. B. Smalley, and A. J. Riker. 1962. Environment and development of Dutch elm disease. *Phytopath.* 52(11):1191–96.

Lanier, G. N., and J. W. Peacock. 1981. Vectors of the pathogen. In *Compendium of elm diseases,* ed. R. J. Stipes and R. J. Campana. St. Paul, Minn: Am. Phytopath. Soc.

Laut, J. G., N. Oshima, and O. E. Dickens. 1969. The recurrence of Dutch elm disease in Colorado. *Plant Dis. Reptr.* 53:253.

Laut, J. G., and M. E. Shomaker. 1979. *Dutch elm disease—a bibliography* (revised). Ft. Collins, Colo.: Colo. State For. Serv., Colo. State Univ.

Laut, J. G., and T. M. Stieger. 1980. Dutch elm disease—a bibliography: Addendum. Ft. Collins, Colo.: Colo. State For. Serv., Colo. State Univ.

Lester, D. T. 1978. Control tactics in research and practice: V. Exploiting host variation. In *Dutch elm disease: Perspectives after 60 years,* ed. W. A. Sinclair and R. J. Campana, 39–42. *Search (Agric.)* 8 (5):1–52.

Liming, O. N. 1937. The Dutch elm disease eradication program—objectives, methods, and results. *Plant Dis. Reptr. Suppl.* 99:18–25.

May, C. 1930. Dutch elm disease in Ohio. *Science* n.s. 72:142–43.

Partridge, A. D., and L. C. Weir. 1974. Dutch elm disease moves into Oregon from Idaho. *Plant Dis. Reptr.* 58:75–76.

Peacock, J. W. 1973. Research on chemical and biological control of elm bark beetles. *Proc.* IUFRO *Conf.,* Sept., 18–53.

Rankin, W. H. 1937. Summary of the Dutch elm disease eradication project. *Proc. Nat. (U.S.) Shade Tree Conf., 1936,* 14–21.

Rankin, W. H., K. G. Park, and D. L. Collins. 1941. Dutch elm disease fungus prevalent in bark beetle infested elm wood. *J. Econ. Entomol.* 34:548–51.

A Review of Dutch elm disease. 1964. Bi-Monthly Progress Report 20(4):1–8.

Ryan, H. J., M. W. Allen, E. C. Calavan, G. E. Carman, W. A. Harvey, P. S. Messenger, W. Reuther, L. L. Sammet, R. F. Smith, T. I. Storer, and S. Wilhelm. 1969. Plant quarantines in California (committee report). Univ. of Calif. Div. of Agric. Sci.

Sargent, C. S. 1961. *Manual of the trees of North America.* 2d ed. Vol. 1. N.Y.: Dover.

Shreiber, L. R., and A. M. Townsend. 1976. Variability in aggressiveness, recovery, and cultural characteristics of isolates of *Ceratocystis ulmi. Phytopath.* 66:239–44.

Sinclair, W. A. 1978a. Range, suscepts, losses. In *Dutch elm disease: Perspectives after 60 years,* ed. W. A. Sinclair and R. J. Campana, 6–8. *Search (Agric.)* 8 (5):1–52.

———. 1978b. Epidemiology. In *Dutch elm disease: Perspectives after 60 Years,* W. A. Sinclair and R. J. Campana, 27–30. *Search (Agric.)* 8 (5):1–52.

Sinclair, W. A., and R. J. Campana, eds. 1978. Dutch elm disease: Perspectives after 60 years. Cornell Univ. Agric. Exp. Stn. *Search (Agric.)* 8 (5):1–52.

Sinclair, W. A., and A. O. Larsen. 1980. Localization of Dutch elm disease in 10-yr.-old white elm clones from resistant parents. *Plant Dis. Reptr.* 64:203–5.

Sinclair, W. A., D. S. Welch, K. G. Parker, and L. T. Tyler. 1974. Selection of American elms for resistance to *Ceratocystis ulmi. Plant Dis. Reptr.* 58:784–88.

Stipes, R. J., and R. J. Campana, eds. 1981. *Compendium of elm diseases.* St. Paul, Minn.: Am. Phytopath. Soc.

Thompson, H. E., and J. G. Matthysse. 1972. Role of the native elm bark beetle, *Hylurgopinus rufipes* (Eichh.), in transmission of the Dutch elm disease pathogen, *Ceratocystis ulmi* (Buisman) C. Moreau. *Cornell Univ. Agric. Exp. Sta. Search* 2(1):1–16.

Tidwell, T. E., J. Kiser, and J. Butler. 1980. Dutch elm disease laboratory annual report. Sacramento: Calif. Dept. of Food and Agric.

Tidwell, T. E., and R. J. Sava. 1982. Isolation of *Ceratocystis ulmi* from deep in the annual rings of elms in California. *Plant Dis.* 66(11):1016–18.

U. S. Department of Agriculture (USDA). 1977. Dutch elm disease: Status of the disease, research, and control. Washington, D.C.: USDA Forest Service.

University of California Cooperative Extension. 1976. The elms in California. *Growing Points newsl.* March 1976.

Van Alfen, N. K., and W. E. MacHardy. 1978. Symptoms and host-pathogen interactions. In *Dutch elm disease: Perspectives after 60 years.* ed. W. A. Sinclair and R. J. Campana, 20–25. *Search (Agric.)* 8 (5):1–52.

Wharton, W. P. 1938. Are the elms being saved. *Amer. For.* 44:545–57.

White, R. P. 1935. Progress of Dutch elm disease eradication. *Plant Dis. Reptr.* 19:270–73.

Wilson, C. H. 1975. The long battle against Dutch elm disease. *J. of Arboriculture* 1:107–13.

10

Citrus Blackfly

Kim A. Hoelmer and J. Kenneth Grace

The management of the citrus blackfly provides an excellent case history of both attempts at eradication and successful biological control. The citrus blackfly, *Aleurocanthus woglumi* Ashby (Homoptera: Aleyrodidae), is a sucking insect that feeds on the foliage of citrus species and other hosts. When present in large numbers the blackfly weakens a plant by lowering the nitrogen levels, thereby reducing the quantity and quality of fruit. It is believed to be a native of the Far East, where it was first noticed by western entomologists in 1910, and has spread throughout most of the citrus-growing regions of the world (fig. 10.1).

Although the blackfly disperses relatively slowly under natural conditions, it is easily transported on leaves of its preferred hosts, *Citrus* species, or on the foliage of such alternate hosts as mango (*Mangifera indica* L.). Incipient infestations are not readily detected because of the small size of all developmental stages of the blackfly and its preference for feeding on the undersides of leaves. The worldwide importance of citrus for fruit production and its popularity as an ornamental in urban plantings make the introduction of citrus blackfly a matter of serious concern. Improved transportation methods in the twentieth century have greatly increased the likelihood of such introductions, either on seedlings or excised leaves in shipments of fruit.

Citrus blackfly first appeared in the western hemisphere in Jamaica in 1913 and spread rapidly throughout the Caribbean region and into Central America. Introductions to the United States have occurred several times.

Early infestations in Florida (1934) and Texas (1955) were eradicated successfully with chemical spray applications. Later infestations in these same two states resisted eradication attempts, however, and were brought under on-going control primarily by the release of parasitic wasps. The success of biological control programs in limiting blackfly numbers to nonpest levels is noteworthy.

We thank R. V. Dowell for reviewing an earlier draft of this chapter.

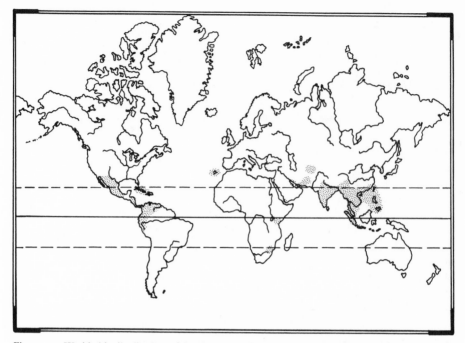

Figure 10.1 Worldwide distribution of the citrus blackfly, *Aleurocanthus woglumi* (adapted from Commonwealth Institute of Entomology 1976).

Life History and Habits

The citrus blackfly, in common with other whiteflies (family Aleyrodidae), has a motile first instar crawler stage, three sessile nymphal instars, and a winged adult stage. Generation time and life history details vary considerably under different environmental conditions (compare Russell 1962; Chavez Torres 1980). Development from egg to adult requires nearly 1000 degrees-days above the developmental threshold of 13.7°C (about two to three months depending on the weather). No development occurs at temperatures below 13.7°C (Dietz and Zetek 1920; Dowell and Fitzpatrick 1978). From two (Afzal Husain and Khan 1945) to six (Dietz and Zetek 1920) overlapping generations per year have been reported under field conditions and a seventh generation is possible under insectary conditions (Clausen and Berry 1932). Three to four generations occur annually in Florida (Dowell et al. 1981).

The female adult citrus blackfly oviposits in a spiral pattern on the underside of leaves. Each female may oviposit several times, with each spiral containing twenty to fifty eggs, for a maximum fecundity exceeding one hundred eggs. Dowell et al. (1981) report the average number of eggs per female to be sixty-five to seventy. Eggs from unmated females develop into male offspring (Dietz and Zetek 1920).

The active first instar nymph, or crawler, is elongate-oval in shape and less than half a millimeter in length. Crawlers usually settle along minor leaf venation within fifteen minutes of emergence (Chavez Torres 1980) but may remain active for several hours (Dowell, Reinert, and Fitzpatrick 1978). Once settled, the insect feeds on plant fluids and grows progressively larger and more ovate through its second, third, and fourth instars. The last nymphal stage is approximately one millimeter in length, and the adult is shorter than two millimeters. Winged adults are initially bright red in color but become covered with a waxy dust within twenty-four hours, producing a general slate-blue appearance (Dowell et al. 1981).

In its native southeast Asian habitat the citrus blackfly is found virtually exclusively on citrus trees (Clausen and Berry 1932). In the western hemisphere, however, *Aleurocanthus woglumi* has been reported on seventy-five different hosts representing thirty families in Cuba, Jamaica, and Panama (Dietz and Zetek 1920) and on at least seventy-five plant species from thirty-eight families in Mexico (Smith, Maltby, and Jimenez 1964). A complete list of hosts on which oviposition has been recorded includes at least one hundred and sixty species in eighty families (Dowell and Steinberg 1979). Female blackflies, like other species of whitefly, apparently orient to plant foliage on the basis of color, responding to reflected light in the 500 to 600 nanometer range (Dowell 1979b), although there is no evidence for long-range chemical attraction. The number of plant species oviposited upon becomes greater with increasing blackfly density (Howard and Neel 1977; Dowell et al. 1979).

Despite the polyphagous oviposition habits of the citrus blackfly, complete development to the adult stage has been reported on only twenty noncitrus species (Dowell et al. 1981), and survival on these alternate hosts is generally low (Clausen and Berry 1932; Smith, Maltby, and Jimenez 1964; Howard and Neel 1977; Dowell and Steinberg 1979; Dowell et al. 1979). Dowell (1978) demonstrated that secondary hosts, including mango (considered the most important), are incapable of supporting citrus blackfly populations without constant immigration from nearby citrus trees.

Biologically and economically, *Citrus* species are the most important hosts, with survivorship greatest and population growth most rapid on lemon, much less on grapefruit, and somewhat intermediate on lime (*Citrus aurantifolia* [Christm.]), orange (*Citrus sinensis* [L.]), tangerine (*Citrus reticulata* Blanco), and tangelo (*Citrus paradisi* Macfad. x *C. reticulata*) (Dowell, Reinert, and Fitzpatrick 1978; Howard 1979). Those noncitrus hosts that are capable of supporting citrus blackfly development from the egg to the adult stage may facilitate spread of the infestation between the more preferred citrus hosts.

Damage and Epidemiology

Feeding by immature citrus blackflies can inflict both direct feeding injury and indirect damage by promoting sooty mold growth. Although there is no evidence that toxins are injected during feeding, insertion of the feeding stylets damages the epidermal cells on the underside of the leaves, resulting in thickened cell walls, loss of cellular contents, and chlorotic patches (Hart, Gausman, and Rodriguez 1976). This cellular damage may be relatively insignificant (Dowell et al. 1981) unless considered in combination with other factors, such as loss of nutrients from the infested tree. Fifty to one hundred blackfly nymphs per leaf can reduce the nitrogen content of that leaf below the 2.2 percent level required by orange trees for successful fruit set (ibid; Dowell 1983).

Indirect damage from citrus blackfly infestation results from the secretion of honeydew by feeding nymphs. Accumulation of honeydew on the leaf surface promotes the growth of sooty mold fungi, impairing respiration and photosynthesis. Damage to leaf cells, nutrient loss, and sooty mold accumulation combine to produce the symptoms characteristic of severe citrus blackfly infestation: reduction in bloom and fruit set, stunting, and defoliation. Branch death or tree kill can occur in severe infestations but is generally uncommon (Ba-Angood 1977; Afzal Husain and Khan 1945). These symptoms can lead to a reduction in fruit yield and quality, although there is little information available regarding the actual economic impact of the citrus blackfly. Most published figures have been offered without supporting data.

Data on economic impact can be difficult to compile, because citrus blackfly infestation is essentially a debilitating condition rather than a direct threat to the fruit itself. Moreover, equivalent densities of nymphs may be associated with quite different effects on tree health because of local differences in edaphic and climatic conditions.

Observers in Panama (Dietz and Zetek 1920) and Florida (R. V. Dowell, personal communication, 1983) were unable to attribute a single case of tree death or loss of vigor to citrus blackfly infestation. Short-term infestations were found to reduce fruit production by up to 50 percent in Mexico (Smith, Maltby, and Jimenez 1964) and Florida, (R. V. Dowell, personal communication, 1983) and infestations of longer than one year frequently resulted in almost complete crop failure. In Pakistan, typical losses due to citrus blackfly have been estimated at 5 to 10 percent with occasional losses as high as 50 to 60 percent (Abbas, Kahn, and Haque 1955). Lotorto (1978) stated that fruit production may be reduced by up to 80 percent in severe infestations. Reductions of this magnitude were not, however, observed in Panama (Dietz and Zetek 1920).

In Florida, the citrus blackfly has been an urban pest, found on dooryard (urban or ornamental) citrus and in nurseries, where it is currently held at low densities by introduced parasites. It has not become a pest in commercial citrus groves. As a result, no economic thresholds have been established (Fitzpatrick, Cherry, and Dowell 1979). Although the emphasis in urban areas has shifted from eradication to containment with biological control, complete exclusion of the citrus blackfly from commercial citricultures is still considered essential.

During the most recent eradication effort in Florida (1976–79), the potential annual costs of chemically controlling the citrus blackfly if it were to become established in commercial groves was estimated to be from $80 million to over $100 million (Cooper 1978; Blackfly eradication program 1976). Such figures do not take into account existing pest management practices. A more carefully documented estimate by Dowell (1980) places a maximum value of $9,288,000 on the additional chemical treatments that would be required annually in Florida to control an established infection without a biological control program.

The citrus blackfly is limited in its distribution by climatic conditions, available host plant material, and its own dispersal capabilities. Warm, humid conditions increase the developmental rate and the number of annual generations; cool, dry conditions retard development. The temperature threshold for citrus blackfly development is 13.7°C, and optimal survival occurs at 26.6°C (Dowell and Fitzpatrick 1978). Dessication of eggs and nymphs under conditions of low humidity can be a major mortality factor, and such conditions also reduce successful adult emergence from the final nymphal instar (Clausen and Berry 1932; Quezada 1974). Adults may also be destroyed in large numbers by heavy rains and wind (Russell 1962).

Adult citrus blackflies do not appear to move great distances on their own. Dowell et al. (1981) report that adults may be capable of flying four hundred to six hundred meters, but very few were caught by yellow sticky traps placed more than fifty meters from an infested tree. First instar crawlers have been observed to walk a maximum distance of only thirty-eight millimeters from the egg spiral (Dietz and Zetek 1920). Although wind dispersal of the crawlers is certainly conceivable, this has not been documented. The possibility of long-range dispersal of adults by wind has also been suggested (Chavez Torres 1980).

Humans are the primary agents of long-distance dispersal of citrus blackfly. *A. woglumi* is thought to have been originally introduced into Jamaica on infested seedlings from its Asian home; live immatures and adults have also been intercepted on boutonnieres, corsages, and excised leaves in fruit shipments (Newell and Brown 1939; Dowell et al. 1981). However, survivorship is low on excised plant parts, and potted seedlings are probably the major means of long-distance dispersal to new areas. Inspection and quarantine are of prime importance in preventing introductions.

Control

Citrus is grown under adverse conditions in much of the citrus blackfly's range in southeast Asia, with seasonal flooding alternating with periods of water stress. In some of these areas, such as Sri Lanka, citrus is largely a casual crop given little or no care after planting (Burke 1967). In other areas citrus has been an important cultivated crop for many years, or its cultivation has been rapidly expanding. Citrus blackfly has been regarded as a serious pest in some of these regions (Abbas, Kahn, and Haque 1955; Afzal Husain and Khan 1945), but in other portions of its range it is scarce, and when it is found, the level of parasitism by natural enemies is often high (DeBach and Bartlett 1951; Rao 1969). Climatic factors and natural enemies prevent citrus blackfly from attaining high populations levels. Several species of parasitic wasps in the genera *Encarsia* (including *Prospaltella* of earlier authors), *Eretmocerus,* and *Amitus* attack citrus blackfly in different parts of Asia (Clausen and Berry 1932; Russell 1962; Smith, Maltby, and Jimenez 1964).

Because humans are a chief factor in the distribution of this pest, plant quarantine and inspection are major means of limiting its spread. Because the blackfly feeds on the foliage and not on the fruit, primary measures are directed at regulating the shipment of citrus seedlings from nurseries. Such quarantine measures can be regionally effective but may be unevenly applied on a worldwide basis.

In the western hemisphere, the initial establishment and spread of *A. woglumi* was unimpeded by its effective natural enemies, which were left behind in Asia. Early chemical suppression techniques involved the use of various oils and sprays applied as emulsions. A spray consisting of paraffin oil, whale oil soap, and water was widely used in the West Indies and in the first U.S. eradication program in Key West, Florida, in 1934 (Newell and Brown 1939). Formulations containing nicotine were also found to be effective in Panama (Dietz and Zetek 1920) and Jamaica (Gowdey 1921).

When citrus blackfly invaded Mexico, extensive chemical trials revealed that rotenone was very effective, as were the relatively new organophosphates, malathion and parathion. DDT was also effective against the blackfly but had the disadvantage of inducing secondary outbreaks of scale insects (Reinert and Neel 1977). Rotenone and malathion were widely used in Mexico during the 1940s and 1950s; they were replaced by carbophenothion early in the 1960s (Enkerlin 1974; Smith, Maltby, and Jimenez 1964). Insecticide trials during the same period of time in Oman and in India also demonstrated the effectiveness of organophosphates for citrus blackfly control (Reinert and Neel 1977).

In the mid-1950s, rotenone and malathion were used successfully in the eradication of small infestations in the Rio Grande Valley in Texas. Following reinfestation of the same area in 1971, they were replaced by dimethoate (ibid). When citrus blackfly was discovered in 1976 for the second time in Florida, malathion was used initially but was replaced by acephate, then unregistered for such purpose, when the Environmental Protection Agency granted an exemption allowing its use. Acephate requires fewer applications and is less phytotoxic than malathion (Lotorto 1978; Selhime 1980).

Biological control has been extraordinarily successful in dealing with *A. woglumi* virtually everywhere it has been used. Concern over its possible introduction into the United States from the Caribbean islands led the United States Department of Agriculture (USDA) to establish a cooperative program with the Cuban government in 1928 to find and introduce natural enemies of the blackfly. Subsequent exploration in Malaysia resulted in the importation and establishment of *Eretmocerus serius* Silvestri in Cuba, Panama, the Bahamas, and Haiti, where this parasite proved very effective in reducing citrus blackfly populations (Clausen and Berry 1932; Clausen 1978).

As citrus blackfly continued to expand its range in the Americas, *Eretmocerus serius* was successfully imported to combat it in Costa Rica, Barbados, and Jamaica (DeBach 1964). *Eretmocerus* was also brought into Mexico but was less successful there, presumably because of an inability to adapt to the more arid environment (Smith, Maltby, and Jimenez 1964). Further exploration in the Far East resulted in the importation and successful establishment in Mexico of *Amitus hesperidum* Silv., *Encarsia opulenta*, (Silv.), and *Encarsia clypealis* (Silv.) (ibid; Flanders 1969). Augmentation of biological control by *Eretmocerus* with *Encarsia opulenta* has since increased the degree of success in Barbados and Jamaica (Clausen 1978). *Encarsia opulenta* alone was successful when introduced into Venezuela (Chavez Torres 1980). *Amitus hesperidum* was reported to be established in Ecuador (Clausen 1978), but no indications of its success are currently available.

In 1959, citrus blackfly was discovered near Durban, South Africa. Eradication was thought to be impractical because of the large number of dooryard citrus trees in the region, the hilly terrain, extensive plantings of mango (an alternate host), and the general lack of commercial and home treatment for scale pests of citrus. Shipments of *Eretmocerus* from Jamaica were subsequently established in South Africa, and the pest potential of citrus blackfly was reduced significantly (Bedford and Thomas 1965). Biological control using introduced parasites against the blackfly was also found to be effective in the Seychelles (Greathead 1971) and in Kenya. Within several years of the initial releases, sooty mold was virtually eliminated from citrus, and commercial control had been obtained without the use of insecticides (Wheatley 1964).

Parasite rearing facilities established in Mexico for the production of

Amitus hesperidum, Encarsia opulenta, and *Encarsia clypealis* were to be important in the eventual biological control of citrus blackfly in the United States as well as in protecting Mexican citriculture.

History of Eradication Efforts

In August 1934, an infestation of citrus blackfly extending over several city blocks was found on the island of Key West, Florida, by State Plant Board inspectors. Various life stages were found on dooryard citrus and on twenty-two other species of plants and trees (Brown 1937). Rapid inspection of the surrounding islands indicated that the infestation was confined to Key West and probably had been introduced from Cuba (ibid). Ironically, the infestation in Cuba had at the time been brought under excellent control by the introduction of parasites. The State Plant Board had commented the previous year that "Florida's horticultural industries are, by reason of the successful introduction (of parasites), greatly protected against danger of introduction of the blackfly, and if there should be such introduction, there is immediately available an effective control measure" (State Plant Board of Florida 1933).

At a meeting in Key West two days after the initial discovery of citrus blackfly, the State Plant Board and the USDA, with the aid of the Emergency Relief Administration, decided upon an immediate eradication campaign. Eradication measures included (1) suitable regulations providing authority for application of eradication measures, (2) enforcement of a quarantine to prevent the movement of uncertified host plants from Key West, and (3) application of an oil emulsion consisting of two parts fish oil soap, two parts oil, and one part water at twenty-one day intervals (Brown 1937). This spray regime was followed for almost three years, from September 1934 until June of 1937 (with a two month interruption in the spring of 1936), although in the final months only citrus and mango were treated (Newell and Brown 1939). No phytotoxicity was noted even though sufficient material was applied with truck-borne power sprayers to drench all parts of the trees (ibid).

The last citrus blackfly in Key West was found on a single leaf in February 1937. The quarantine was suspended on April 13, 1938, and on April 14, the successful outcome of the eradication campaign was reported at the annual meeting of the Florida State Horticultural Society (Brown 1939). The project was reported to have cost $36,000 in federal funds and $161,464 in state funds (ibid).

The isolation of Key West and the established presence of State Plant Board personnel on the island contributed immensely to the success of the eradication campaign. These same factors, however, also created serious problems during the course of the campaign.

Key West was physically isolated from mainland Florida in the 1930s. The overseas railroad bridge was partially destroyed by a hurricane in 1935, and construction of the automobile causeway was not begun until 1937. The five-hour ferry ride and only three airplane flights each week simplified the enforcement of quarantine measures.

But the citizens of Key West were socially isolated as well from mainland Florida. In 1937, the *New York Times* referred to Key West as "problem stepchild of the United States, neglected by the parent mainland, gone foreign in atmosphere, architecture, habit and language" (Berger 1937). There were no commercial citrus groves on the island, and the State Plant Board found residents to be indifferent towards the threat to mainland groves and hostile to the eradication effort.

Opposition to the spray program was encountered almost immediately among property occupants. The personal inconvenience of the program was certainly a major factor in this resistance, since it was necessary to provide access to the spray crews every three weeks in order to "drench" the twenty-three suspected host plants. In some cases, it was necessary to drag great lengths of hose through residents' homes in order to reach yard plantings (Newell and Brown 1939). The possibility of contaminating back-yard cisterns (rainwater being the sole source of fresh water on Key West) may have been a contributing fear. This limited availability of fresh water caused supply problems for the Plant Board as well, since their crews used several thousand gallons daily (ibid).

A simultaneous (1933–36) eradication campaign against the West Indian fruit fly (*Anastrepha* spp.) on Key West undoubtedly contributed to citizen inconvenience. This project involved fruit stripping and the widespread application of tartar emetic (antimony and potassium tartrate) at fifteen-day intervals to yard plantings. This eradication attempt was dropped in 1936, when it became evident that the West Indian fruit fly had been present for some time on the mainland without any apparent economic damage (Brown 1937). Because the State Plant Board was an established presence in Key West when the citrus blackfly was first detected in 1934, trained person-nel and equipment were already available locally. A somewhat antagonistic relationship with much of the local citizenry also existed.

In May 1935, criminal charges were filed against one objector for violating plant board regulations, but the local jury failed to reach a verdict. Later in 1935 and in 1936, several citizens were found guilty of assaulting plant board employees. In June 1935, the plant board was denied an injunction restrain-ing fifty-five objecting property occupants. However, in December a per-manent injunction was granted by the circuit court in Miami. The defen-dents in the suit appealed to the Florida State Supreme Court, which nullified the restraining order while acting upon the appeal in March 1936. By this time, the number of objectors had increased to approximately one hundred (ibid), and the plant board, "disappointed at its own failure as well as that of interested and public-spirited citizens to reduce opposition, had discontinued spraying and withdrew its inspection force" (Brown 1939).

The interruption in spraying was short-lived. On March 30, 1936, twelve days after operations ceased in Key West, a joint meeting of the State Plant Board and the Florida Citrus Commission was held to advise citrus industry leaders of the situation. Immediately after this meeting, plans were made for resuming operations (Brown 1937).

In May 1936, the supreme court sustained the injunction granted by the circuit court, and spray operations resumed. A deputy sheriff was assigned to each spray truck to protect the crew (ibid). General opposition to the eradication project declined greatly, and only a few individuals were cited for contempt before eradication activities were discontinued in June 1937. Quarantine enforcement was dropped in April 1938. Although this brought the four-year eradication campaign to a successful conclusion (confirmed by an extensive inspection in the fall of 1938), the State Plant Board emphasized that with proper cooperation from property occupants, the operation would have been completed in one year or less (Brown 1937, 1939; Newell and Brown 1939).

In 1947, concern over citrus blackfly infestations in Guaymas and Em-palme, Mexico, 270 miles south of the Arizona border, led California and Arizona citrus growers to contribute $25,000 for a chemical control pro-gram in these two cities and nearby citrus groves. With the cooperation of the Mexican Dirección General de Agricultura all urban host trees and commercial groves in the area were sprayed two to three times with ro-tenone in oil between November 1947 and April 1948. In 1948, responsibility for this grower-initiated program passed to the USDA and the Dirección General de Agricultura, and chemical treatments and monitoring were con-tinued (Cooper, Plummer, and Shaw 1950).

As in Key West, spray hoses frequently had to be dragged through resi-dents' homes in Guaymas and Empalme in order to reach enclosed patios with trees (Gunter 1954). No enforcement laws existed, and compliance was at the discretion of the individual property owners. Initially, the urban public was reported to be quite cooperative (Woglum 1948). But several years and many treatments later the lack of enforcement laws was cited as a major stumbling block to eradication of the persistent low-level infestations of blackfly. Where spraying was not permitted, labor-intensive hand strip-ping of infested leaves was used instead (Woglum, Smith, and Clausen 1952; Gunter 1954).

In early 1950, a small blackfly infestation was found in two trees in Matamoros, Mexico, immediately across the border from Brownsville,

Texas. All host trees within a nine-block square area were sprayed four times with rotenone in oil (Berry 1951). The continuing threat of insect encroachments across the Texas border and into the citrus groves of the Lower Rio Grande Valley resulted in the formation of the Mexican Fruit Fly-Citrus Blackfly Control Project in 1951 (Hart et al. 1973). Personnel of the state of Texas, the USDA, and the Dirección General de Agricultura conducted ground surveys, treated and eradicated spot infestations along the Texas-Mexico border, and maintained a blackfly-free barrier zone in northwestern Mexico (Reinert and Neel 1977). South of the barrier zone blackfly was effectively controlled by parasites.

During 1955–56 citrus blackfly began to appear in Texas. By the end of 1956 thirty-three separate infestations had been reported (ibid). These were small infestations on leaves attached to fruit or seedlings brought across the border or small infestations on noncommercial trees growing near the border. The latter were believed to be due to tourists bringing adult blackflies across the border in cars or trailers. A series of localized malathion or rotenone sprays were successful in eliminating these foci.

New blackfly infestations in Texas were reported during 1967 and were attributed to dispersal from Mexico by hurricane winds originating from the Yucatan peninsula six to seven months earlier (Chavez Torres 1980). By 1971, citrus blackfly invasions of Texas had triggered another eradication program, this time with dimethoate sprays applied every three weeks (Hart et al. 1978). Despite repeated applications, with some trees receiving up to thirty treatments, the infestation not only persisted but continued to spread, moving beyond dooryard citrus into several commercial groves. To help combat further spread, federal quarantine regulations were instituted requiring all commercial citrus to be stripped of leaves prior to being packed for shipment. Citrus blackfly was successfully confined to residential citrus in the Brownsville area until 1974 when it moved into commercial groves. A program to treat infested groves with dimethoate was immediately instituted (Citrus blackfly fight 1975).

In 1974, a trial area of heavily infested dooryard citrus was chosen to test the effectiveness of biological control, while the eradication program was continued outside of this zone. Field-collected *Encarsia opulenta* and *Encarsia clypealis* and laboratory-reared *Amitus hesperidum* were released onto residential citrus in Brownsville (Holler and Brazzel 1978). After three years, blackfly populations had been reduced to low levels, and no commercial production losses were reported during the final two years of the study (Ketner and Rosier 1978). The most recent studies have shown that citrus blackfly in the Lower Rio Grande Valley continues to be held at very low population densities by the action of *Encarsia opulenta* (Summy et al. 1983).

At about the same time that *A. woglumi* was coming under satisfactory biological control in Texas, it was discovered anew in Florida. In January

1976, citrus blackfly was found on nursery stock in an urban area of Fort Lauderdale, Broward County. An initial survey of the surrounding area indicated that the infestation covered a minimum of several square miles, and spraying was promptly scheduled (FDACS 1976a). Further surveys were conducted, and within a month the size of the area placed under quarantine had expanded to two hundred square miles. Because the blackfly disperses slowly, the size of the Fort Lauderdale infestation was believed to indicate that the insect had been in Florida for at least four years—possibly even longer—prior to its detection (Hart et al. 1978; R. V. Dowell, personal communication, 1983).

Quarantine regulations went into effect in March 1976, requiring that nurseries in the affected areas be inspected on a regular basis and that all mango and citrus plants transported from the controlled areas be treated and/or certified to be free of citrus blackfly. Fruit shipped from this region was to be fumigated with methyl bromide or certified to be free of all leaves. An official dumping site was designated for the disposal of yard trash and plant clippings.

An urban biometric survey was initiated at this time to provide detailed information on the size and distribution of the infestation (FDACS 1976b). Initially, the USDA contracted with commercial applicators to conduct door-to-door spraying with malathion in the infested portion of Broward County (Blackfly eradication program 1976; FDACS 1976a).

The success of biological control in dealing with citrus blackfly elsewhere and the difficulty in eradicating the pest from Texas led some researchers to conclude that biological control was a preferable method of dealing with the situation. Citrus industry representatives disagreed, however, believing that climatic conditions in Florida were sufficiently different from other citrus growing regions to permit eradication (Hardy 1976; R. V. Dowell, personal communication, 1983). Within two weeks of the initial blackfly discovery in Fort Lauderdale, the board of directors of Florida Citrus Mutual requested that the state department of agriculture and the USDA immediately institute a chemical eradication program (Reitz 1977). As planning for an eradication campaign proceeded, however, arrangements were made to import parasites from rearing facilities in Mexico as part of an integrated management study intended partly as a hedge against an unsuccessful eradication attempt (Dowell 1979a). Releases of *Amitus hesperidum* and *Encarsia opulenta* were begun in April 1976. A third species, *Encarsia clypealis,* was released once at a single location in November but was not recovered again (Hart et al. 1978).

Meanwhile, the size of the infestation continued to grow in proportion to the area examined in the biometric survey. By mid-1976 the regulated area had increased to 480 square miles. In July the state legislature officially authorized a cooperative eradication program to be administered jointly by the USDA Animal and Plant Health and Inspection Service and the Florida Division of Plant Industry. This effort was projected to last five years and cost about five million dollars per year; six dollars per year for every acre of commercial citrus in the state (Lotorto 1978; Reitz 1977).

At the completion of the blackfly survey in August 1976, the infestation covered the southern half of Palm Beach County, most of the inhabited portions of Broward County, and the northern half of Dade County. Both state and federal quarantines were designed to prevent the citrus blackfly from spreading north and west from these urban areas in southeastern Florida to the primary citrus growing regions in central Florida.

Chemical control was initiated in August using six applications of malathion, applied by ground crews at two-week intervals. Sprays were applied in two moving buffer zones, beginning at the northern and southern boundaries of the eradication zone and moving towards each other. In October an Environmental Protection Agency exemption authorized limited use of acephate in place of malathion. Acephate had been found to have both residual and systemic toxicity against the blackfly and was considered to be less phytotoxic and less hazardous to birds and fish than malathion. The numbers of sprays applied per property in the buffer zones was reduced to three with acephate instead of the previous six (Lotorto 1978; Reinert and Neel 1977). Initial hopes for the development of aerial application methods (FDACS 1976b) were never realized. Reitz (1977) attributed this to the preference of the blackfly for the undersides of leaves and to insufficient translocation within plants of the pesticides employed.

An extensive, well-organized media campaign was mounted by the state of Florida to inform the general public about the citrus blackfly and the eradication program (Dowell et al. 1979; Reitz 1977). Perhaps as a result, the blackfly campaign did not experience the negative public reception in the affected areas that occurred in 1934 in Key West.

The state legislature also required basic research on the biology of the blackfly and its chemical and biological control as part of the eradication program. Researchers at the University of Florida cooperated with state and federal workers in producing new information on the biology and host range of *A. woglumi,* efficacy of various control measures, and biology of its parasites (Lotorto 1978; Reinert and Neel 1977; R. V. Dowell, personal communication, 1983).

Funding of the eradication program raised some interesting questions. A state legislature proposal to finance an emergency pest control fund by levying a new tax on agricultural commodities was strongly opposed by agricultural interests on the grounds that additional industry taxes were inappropriate for solving pest problems that may have been caused by the general public (Lavigne 1976). The citrus industry also opposed grower-financed projects to raise money for the eradication program (Hardy 1976).

This opposition by industry succeeded in overriding the tax proposal, and the eradication program was funded from general revenue moneys.

Continued monitoring during 1977 indicated that the infestation centered in Fort Lauderdale actually encompassed 1,250 square miles (Lotorto 1978). The two buffer zones to the north and south of the infested area continued to move inward; affected citrus was treated with acephate every twenty-one days. In October 1977, mango and Surinam cherry were added to the treatment schedule. The addition of these two hosts, together with the increased size of the area to be treated, made continued treatment of both barrier zones economically unfeasible. Therefore, treatment of the southern zone was discontinued in early 1978, and all effort was concentrated in the northern part of the infested area nearest the important citrus-growing regions to the northwest.

Parasite releases were also continued throughout 1977, with great success in establishment and in controlling citrus blackfly. By January 1978, the parasites alone were credited with a 97 percent reduction in the blackfly population (FDACS 1978; Selhime 1979). This led to the incorporation of biological control as an integral part of the eradication effort: parasite releases were used to rapidly reduce blackfly numbers with subsequent acephate applications intended to eliminate the remainder. The opportunity to capitalize on the blackfly reduction resulting from parasites in the northern portion of the infested area was offered as an additional rationale for concentrating the spray operations in that region (Florida expands 1978).

Surveys in late 1978 and early 1979 revealed a persistent low-level infestation throughout the sprayed areas rather than the expected spot infestations which would have indicated local reintroductions (R. V. Dowell, personal communication, 1983). The two parasites, *Encarsia opulenta* and *Amitus hesperidum,* also managed to survive the spray regime and persisted at very low host densities (Dowell 1979a; Cherry and Pastor 1980).

The proven effectiveness of the parasites and their persistence at low densities of the citrus blackfly led the Blackfly Technical Advisory Committee to recommend termination of the eradication phase and full implementation of a biological control program. Despite disagreement by some associates of the citrus industry (Blackfly question 1979), Florida Commissioner of Agriculture Doyle Conner announced in March 1979 his decision to accept a "containment" program with the option of reverting to eradication if necessary (FDACS 1979). The eradication program had not officially failed but simply was no longer necessary.

The containment program provided for continued monitoring of citrus blackfly populations in the area of original infestation, parasite releases, and chemical treatment of isolated infestations. All treatment with acephate was terminated in September 1979 when the specific exemption for its use expired. As of 1988 citrus blackfly is considered under complete biological control in Florida. Occasional local flareups are quickly eliminated by *Amitus*. Both parasite species have become established and are no longer being reared for release (R. V. Dowell, personal communication, 1988).

Thus, the speculation in 1933 by the Florida State Plant Board that parasites could effectively control citrus blackfly in mainland Florida had finally been accepted, though not without two attempts at eradication. The first attempt, in 1933, encountered a great deal of public resistance but proved successful because of the isolated nature of the infested area (Key West). Forty years later, the second eradication effort aroused no public antagonism yet failed to attain its goal. Despite public support for the program, the size of the infested area and its location in a major urban center undoubtedly made thorough treatment and enforcement of quarantines difficult. With citrus blackfly present in Mexico and the Caribbean, the establishment of effective parasites in Florida and Texas serves not only to control existing populations but also as insurance against the ever-present threat of reintroduction.

Evaluation of Eradication Efforts

Successful eradications of *Aleurocanthus woglumi* share several comparable features. In each case the total area of infestation was very small or very isolated at the time of detection. Survey or monitoring techniques used were very extensive or were already established for the blackfly or for other pests. Failure to eradicate citrus blackfly, therefore, was often attributable to late detection. Complicating this was the tendency of the insect to appear first in an urban setting where survey and control measures are labor inten-sive and thus inherently expensive and less efficient. The resistant nature of the immature stages and their habit of feeding on the undersides of leaves also contributed to the lack of effectiveness of control methods.

Prevention of pest establishment with rapid chemical treatment is often a cost-effective means of dealing with a new introduction if the pest can be detected and eliminated before it spreads throughout a new habitat. With the citrus blackfly, the requirement of early detection has seldom been met and will probably remain a problem in the future.

The continual growth of worldwide travel and trade seems likely to assure new introductions and reintroductions of *A. woglumi*. Localized spot treat-ments of the pest are desirable in attempting to prevent establishment. But the failures of previous large-scale eradication programs should be kept in mind. Biological control remains an effective management technique, and recent research illustrating the compatibility of natural enemies with the pest management techniques currently practices in citriculture (Fitzpatrick, Cherry, and Dowell 1979; Dowell and Fitzpatrick 1980) should strengthen the acceptance of biological control by the citrus industry. Natural control of citrus blackfly offers a textbook example of effective classical biological control.

Literature Cited

Abbas, H. M., M. S. Kahn, and H. Haque. 1955. Blackfly of citrus (*Aleurocanthus woglumi* Ashby) in Sind and its control. *Agric. Pakistan* 6(1):5–23.

Afzal Husain, M., and Abdul Wahid Khan. 1945. The Citrus Aleurodidae (Homoptera) in the Punjab and their control. *Mem. Entomol. Soc. India*, no. 1:26–29.

Ba-Angood, S. A. S. 1977. Field trials for the control of *Aleurocanthus woglumi* in the Yemen. *PANS* 23:149–52.

Bedford, E. C. B., and E. D. Thomas. 1965. Biological control of the citrus blackfly *Aleurocanthus woglumi* (Ashby) (Homoptera: Aleyrodidae) in South Africa. *J. Entomol. Soc. S. Afr.* 28(1):117–32.

Berger, M. 1937. Old Key West awakes. *New York Times Magazine*, March 21.

Berry, N. O. 1951. Progress report on the citrus blackfly surveys in Mexico. *Proc. Rio Grande Valley Hort. Soc.* 5:81–82.

Blackfly eradication program under way in Broward county. 1976. *Fla. Grower and Rancher* 68:19–20.

The blackfly question: Eradicate or biocontrol? 1979. *Fla. Grower and Rancher* 72(4):12–13.

Brown, A. C. 1937. Report of the grove inspection department. *Eleventh Biennial Report (1934–1936)*, 15–27. Gainesville: State Plant Board of Florida.

———. 1939. Report of the grove inspection department. *Twelfth Biennial Report (1936–1938)*, 15–17. Gainesville: State Plant Board of Florida.

Burke, J. H. 1967. The commercial citrus regions of the world. In *The citrus industry*, Vol. 1, *History, world distribution, botany, and varieties*, ed. W. Reuther, H. J. Webber, and L. D. Batchelor, 40–189. Berkeley: Univ. Calif. Div. Agric. Sci.

Chavez Torres, H. A. 1980. Biological control of the citrus blackfly (*Aleurocanthus woglumi* Ashby) by *Prospaltella opulenta* Silv. in central-western Venezuela with a review of the pest's invasion of the western hemisphere and suppression by introduced parasites. Ph.D. diss. Univ. of Florida, Gainesville.

Cherry, R., and S. Pastor, Jr. 1980. Variations in population levels of citrus blackfly, *Aleurocanthus woglumi* (Hom.: Aleyrodidae) and parasites during an eradication program in Florida. *Entomophaga* 25:365–68.

The citrus blackfly fight in the lower Rio Grande Valley. 1975. *Citrus Ind.* 56(7):8,10–11.

Citrus blackfly: Florida eyes eradication. 1976. *Calif. Citrogr.* 62(2):43–44.

Clausen, C. P. 1978. Introduced parasites and predators of arthropod pests and weeds. *U.S. Dept. Agric. Handbk.* 480, 30–34.

Clausen, C. P., and P. A. Berry. 1932. The citrus blackfly in Asia, and the importation of its natural enemies into tropical America. *U.S. Dept. Agric. Tech. Bull.* 320.

Commonwealth Institute of Entomology. 1976. *Distribution maps of pests.* Map no. 91. London: Commonwealth Institute of Entomology.

Cooper, J. F. 1978. Eradication—An objective forcefully pursued. *Fla. Grower and Rancher* 71:14–16.

Cooper, J. F., C. C. Plummer, and J. G. Shaw. 1950. The citrus blackfly situation in Mexico. *J. Econ. Entom.* 43:767–73.

DeBach, P. 1964. Successes, trends, and future possibilities. In *Biological control of insect pests and weeds,* ed. P. DeBach, 673–713. N.Y.: Reinhold.

DeBach, P., and B. Bartlett. 1951. Effects of insecticides on biological control of insect pests of citrus. *J. Econ. Entom.* 44:372–83.

Dietz, H. F., and J. Zetek. 1920. The blackfly of citrus and other tropical plants. *U.S. Dept. Agric. Tech. Bull.* 885.

Dowell, R. V. 1978. Suitability of mango as a long-term host of the citrus blackfly. *Proc. Fla. State Hort. Soc.* 90:229–30.

———. 1979a. Synchrony and impact of *Amitus hesperidum* (Hym.: Platygasteridae) on its host, *Aleurocanthus woglumi* (Hom.: Aleyrodidae) in southern Florida. *Entomophaga* 24:221–27.

———. 1979b. Host selection by the citrus blackfly *Aleurocanthus woglumi* (Homoptera: Aleyrodidae). *Ent. Exp. and Appl.* 25:289–96.

———. 1980. Economics of biological control of blackfly. *Citrus Ind.* 61(12):12–16.

———. 1983. Nitrogen levels in citrus leaves infested with immature citrus blackfly. *Ent. Exp. and Appl.* 34:201–3.

Dowell, R. V., R. H. Cherry, G. E. Fitzpatrick, J. A. Reinert, and J. L. Knapp. 1981. Biology, plant-insect relations, and control of the citrus blackfly. *Agric. Exp. Sta., Inst. Food and Agric. Sci., Univ. Fla., Bull.* 818.

Dowell, R. V., and G. E. Fitzpatrick. 1978. Effect of temperature on the growth and survival of the citrus blackfly. *Can. Entomol.* 110:1347–50.

———. 1980. Citrus blackfly: Influence of spray on its biological control. *Citrus Ind.* 61:29–30, 33, 36.

Dowell, R. V., F. W. Howard, R. H. Cherry, and G. E. Fitzpatrick. 1979. Field studies of the host range of the citrus blackfly *Aleurocanthus woglumi* Ashby (Homoptera: Aleyrodidae). *Can. Entomol.* 111:1–6.

Dowell, R. V., J. A. Reinert, and G. E. Fitzpatrick. 1978. Development and survivorship of the citrus blackfly, *Aleurocanthus woglumi,* on six citrus hosts. *Environ. Entomol.* 7:524–25.

Dowell, R. V., and B. Steinberg. 1979. Development and survival of immature citrus blackfly (Homoptera: Aleyrodidae) on twenty-three plant species. *Ann. Entomol. Soc. Amer.* 72:721–24.

Enkerlin, S. D. 1974. Some aspects of the citrus blackfly (*Aleurocanthus woglumi* Ashby) in Mexico. *Tall Timbers Conf. Ecol. Anim. Control Habitat Management,* no. 6, 65–76.

FDACS. 1976a. *Tri-ology technical report* 15(1), Fla. Dept. Agric. and Cons. Serv. (*FDACS*). Tallahassee: Div. Plant Ind. (Bur. Entomol.).

FDACS. 1976b. *Tri-ology technical report* 15(3). See *FDACS* 1976a.

FDACS. 1978. *Tri-ology technical report* 17(1). See *FDACS* 1976a.

FDACS. 1979. *Tri-ology technical report* 18(3). See *FDACS* 1976a.

Fitzpatrick, G. E., R. H. Cherry, and R. V. Dowell. 1979. Effects of Florida citrus pest control practices on the citrus blackfly (Homoptera: Aleyrodidae) and its associated natural enemies. *Can.Entomol.* 111:731–34.

Flanders, S. W. 1969. Herbert D. Smith's observations on citrus blackfly parasites in India and Mexico and the correlated circumstances. *Can. Entomol.* 101:467–80.

Florida expands citrus blackfly work. 1978. *Calif. Citrogr.* 63(5):121,124.

Gowdey, C. C. 1921. The citrus blackfly (*Aleurocanthus woglumi* Ashby). *Jamaica Dept. Agric. Entomol. Circ.* 3.

Greathead, D. J. 1971. A review of biological control in the Ethiopian region. *Commonwealth Inst. Biol. Contr. Tech. Comm.* 5.

Gunter, F. S. 1954. Black fly control in Mexican groves. *Citrus Leaves* 34:20.

Hardy, N. G. 1976. Citrus comments. *Fla. Grower and Rancher* 69:19.

Hart, W. G., H. W. Gausman, and R. R. Rodriguez. 1976. Citrus blackfly (Hemiptera: Aleyrodidae), feeding injury and its influence on the spectral properties of citrus foliage. *J. Rio Grande Valley Hort. Soc.* 30:37–43.

Hart, W. G., S. J. Ingle, M. R. Davis, and C. Mangum. 1973. Aerial photography with infrared color film as a method of surveying for citrus blackfly. *J. Econ. Entomol.* 66:190–94.

Hart, W. G., A. Selhime, D. P. Harlan, S. J. Ingle, R. M. Sanchez, R. H. Rhode, C. A. Garcia, J. Caballero, and R. L. Garcia. 1978. The introduction and establishment of parasites of citrus blackfly, *Aleurocanthus woglumi* in Florida (Hom.: Aleyrodidae). *Entomophaga* 23:361–66.

Holler, T. C., and J. R. Brazzel. 1978. Colonization of citrus blackfly parasites in the lower Rio Grande Valley of Texas. *Citrus Ind.* 59:42–46.

Howard, F. W. 1979. Comparacion de seis especies de *Citrus* como plantas hospederes de *Aleurocanthus woglumi* Ashby. *Folia Entomol. Mex.* 41:57–64.

Howard, F. W., and P. L. Neel. 1977. Host plant preferences of citrus blackfly *Aleurocanthus woglumi* Ashby (Homoptera: Aleyrodidae), in Florida. *Proc. Int. Soc. Citriculture* 2:489–92.

Ketner, C. F., and J. G. Rosier. 1978. Citrus blackfly controlled biologically. *Texas Agric. Prog.* 24:19–20.

Lavigne, N. F. 1976. "Emergency" ag tax killed. *Fla. Grower and Rancher* 69(9):14.

Lotorto, G. F. 1978. Eradication of citrus blackfly, biological and chemical control. *Proc. Fla. State Hort. Soc.* 91:192–93.

Newell, W., and A. C. Brown. 1939. Eradication of the citrus blackfly in Key West, Florida. *J. Econ. Entomol.* 32:680–82.

Quezada, J. R. 1974. Biological control of *Aleurocanthus woglumi* (Homoptera: Aleyrodidae) in El Salvador. *Entomophaga* 19:243–54.

Rao, V. P. 1969. India as a source of natural enemies of pests of citrus. *Proc. First Int. Citrus Symp.* 2:785–92.

Reinert, J. A., and P. L. Neel. 1977. A history of chemical control for the citrus blackfly and development of its management in Florida. *Proc. Int. Soc. Citriculture* 2:493–96.

Reitz, H. J. 1977. Prospects for the future of citrus blackfly in Florida. *Proc. Fla. State Hort. Soc.* 90:14–16.

Russell, L. M. 1962. The citrus blackfly. *FAO Plant Prot. Bull.* 10:36–38.

Selhime, A. G. 1979. Biological control of citrus blackfly in South Florida. *Proc. Fla. State Hort. Soc.* 92:32–33.

————. 1980. Selhime responds to blackfly series. *Citrus Ind.* 61:16–20.

Smith, H. D., H. L. Maltby, and E. J. Jimenez. 1964. Biological control of the citrus blackfly in Mexico. *U.S. Dept. Agric. Tech. Bull.* 1311.

State Plant Board of Florida. 1933. Parasitic control of blackfly in Cuba. *Ninth Biennial Report (1930–1932)*:21–22. Gainesville: State Plant Board of Florida.

Summy, K. R., F. E. Gilstrap, W. G. Hart, J. M. Caballero, and L. Saenz. 1983. Biological control of the citrus blackfly in Texas. *Environ. Entomol.* 12:782–86.

Wheatley, P. E. 1964. The successful establishment of *Eretmocerus serius* Silv. (Hymenoptera: Eulophidae) in Kenya. *East Afr. Agric. For. J.* 29:236.

Woglum, R. S. 1948. The citrus black fly campaign in Sonora gets under way. *Calif. Citrogr.* 33:118–19.

Woglum, R. S., H. S. Smith, and C. P. Clausen. 1952. The citrus blackfly in Northwest Mexico. *Calif. Citrogr.* 37:356.

II

Oriental Fruit Fly

John S. Yaninek and Francis Geraud

A single oriental fruit fly, *Dacus dorsalis* Hendel, was discovered in May 1946 in a trap collected during November 1945 on the island of Hawaii. A reexamination of previous trap catches revealed that other oriental fruit flies had been found earlier in the year on Oahu. The introduction of the fly was estimated to have occurred prior to 1945, probably a consequence of troop movements throughout the western Pacific earlier in the decade. By 1946 this fly was found on every major island in the territory.

Confirmation of this new pest represented the third fruit fly invasion of the islands in fewer than fifty years. Two previous introductions—the melon fly (*Dacus cucurbitae* Coq.) in 1897 and the Mediterranean fruit fly (*Ceratitis capitata* [Weid.]) in 1910—were already responsible for considerable losses to local growers. The addition of the oriental fruit fly loomed as a fatal blow to a fledgling export industry (Carter 1950). Territorial merchants and agricultural officials agreed that prompt, decisive action was needed. In 1949, a representative of the Honolulu Chamber of Commerce proclaimed this latest fruit fly infestation to be an emergency of national proportions and asked Congress for $2 million to help fight the pest.

Initially, the Territorial Board of Agriculture and Forestry was responsible for the oriental fruit fly problem, but by 1948 the problem was rapidly growing beyond the financial means of the territorial government and the technical capacity of its agencies. Four additional agencies joined the efforts of the Territorial Board of Agriculture and Forestry. These included two private research units, the Hawaiian Sugar Planters' Association and the Pineapple Research Institute, and two government units, the Hawaii Agricultural Experiment Station and the Bureau of Entomology and Plant Quarantine, which eventually assumed overall responsibility for coordinating research activities between agencies.

Territorial farmers, researchers, and administrators were immediately aware of the problems this new pest represented. But their counterparts on the U.S. mainland were slower to respond (Armitage 1949). The local pest situation became so serious by 1948 that produce in the marketplace, usually free of adult fruit flies, was heavily attacked. Oriental fruit flies began showing up in produce being shipped to the west coast of the United States. Word of the ability of the fly to attack more than 250 species of economic plants, the capacity for ten generations per year, and a vigor greater than that of the Mediterranean fruit fly finally reached California, and in 1948 a legislative delegation was sent to Hawaii to investigate the situation.

Financial and technical cooperation soon followed. The primary reason for this was the threat the fly represented to mainland agriculture; consequently much of the research effort was designed with this in mind. Congress responded by appropriating a total of $2 million, to be spread over four years (1949–52).

There was considerable interest in this research effort in California, where agriculture was and is the number one state industry. California was also the first state likely to be invaded by the fruit fly. The state legislature appropriated $800,000 over the same four-year period to develop monitoring techniques and eradication strategies in case the oriental fruit fly populations in Hawaii were not eradicated. Federal and state contributions, including the $375,000 spent by the territory, and the $40,000 spent on four bioclimatic chambers brought the total budget of the project to more than $3 million. In short, the project had a realistic budget for dealing with a serious pest.

Taxonomy, Life History, and Damage

The oriental fruit fly, *Dacus (Strumeta) dorsalis* Hendel, belongs to the subfamily Dacinae of the family Tephritidae (Diptera). This subfamily is largely confined to tropical and subtropical regions of Asia and Australia. According to Clausen (1977), the geographic distribution of the oriental fruit fly consists of the Indo-Malayan region, with northward extensions into south China and many islands of the western Pacific (fig. 11.1). It was originally described under the name of *Musca ferruginea* Fabricius (1794), thus becoming a junior homonym of the prior *Musca ferruginea* Scopoli (1763) (Hardy 1969). It was renamed by Hendel in 1912. This insect has also been mentioned in the literature as *Batrocera conformis* Doleschall and *Dacus ferrugineus* var. *mangiferae* Cotes.

Female oriental fruit flies mate and oviposit eight to twelve days after the adult emergence, depending on nutritional background. As with several other fruit flies, the rate of sexual maturity and the eventual fecundity depends on the availability of essential nutrients during the postemergence period (Hagen 1958). Eggs are deposited in clusters beneath the skin of ripening fruits by insertion of the terminal three segments of the ovitube,

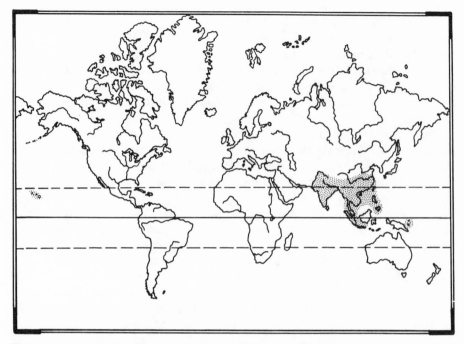

Figure 11.1 Geographical distribution of the oriental fruit fly *Dacus dorsalis* (adapted from International Institute of Entomology 1986).

not a true ovipositor in flies. Eggs can be deposited in the oviposition wound made by other females and other insect species, as well as in depressions and cracks of the fruit skin. The newly hatched larvae bore into the fruit and molt twice while feeding. Once the damaged fruit falls to the ground, the third instar larvae leave and burrow two to three cm into the soil for pupation (Christenson and Foote 1960). Length of development of the larvae to adult varies from about twenty days to three months, depending on the climatic conditions. Adults can live from one to three months, depending on climatic and nutritional conditions.

The oriental fruit fly has been found attacking more than 150 species of fruit and vegetables, among which mango (*Mangifera indica* L.), guava (*Psidiam guajava* L.), citrus (*Citrus* spp., especially loose-skinned species), avocado (*Persea americana* Mill.), papaya (*Carica papaya* L.), and tomato (*Lycopersicum sculentum* Mill.) are especially commercially important (Clausen 1977). Syed, Ghani, and Martaza (1970) also mention watermelon (*Citrullus vulgand* Schrad.), several species of plums and peaches (*Prunus* spp.), several species of pears (*Pyrus* spp.), and one of apple (*Malus pumila*), among others, as host plants in West Pakistan.

Epidemiology and Ecology

Like other tropical and subtropical fruit flies, the oriental fruit fly is a multivoltine species, and it is not known to undergo diapause (Christenson and Foote 1960). Under inclement climatic conditions, it congregates in locations that provide shelter and food (Bateman 1972). Individuals of both sexes can mate several times during their lives. A single mating ensures a female fertility for life, although maximum fertility requires multiple matings. Syed, Ghani, and Martaza (1970) reported an average reproductive output of 720 eggs per female with a maximum of 82 eggs in a day under laboratory conditions. The risk of a mainland invasion, particularly in California, initiated an intensive study of the climatic limits of the oriental fruit fly. Reproductive biology and survivorship were examined in bioclimatic chambers adjusted to reflect potentially favorable conditions throughout the United States (Messenger and Flitters 1954; Messenger 1960). From the eighteen climatic patterns included in this study, continuous favorable conditions were found in southern Florida, Louisiana, and Texas (Messenger and Flitters 1954), as well as a narrow coastal plain between Laguna Beach and Oceanside in southern California (Messenger 1960). Average monthly temperatures of less than 13.9°C (57°F) and more than 32.2°C (90°F) over a three-month period or an average daily temperature exceeding 37.8°C (100°F) seemed to be limiting conditions for continuous development (Messenger and Flitter 1954).

Generally speaking, tropical fruit flies are strong fliers. These flies tend, however, to display nondispersion movements in climatically favorable areas as long as suitable feeding, mating, and oviposition resources are available (Bateman 1972). Adult male oriental fruit flies have been reported to cover distances of up to 50 km between some of the southern Japanese islands (Iwahashi 1972). This capacity to travel long distances allows the fly to disperse from poor areas and colonize new and potentially more promising sites.

Strong competitive interaction seems to have occurred in Hawaii between the already existing Mediterranean fruit fly, *Ceratitis capitata* [Weid.], and the more recently introduced oriental fruit fly (Bess 1953). As a result, *C. capitata* now has a more restricted geographic and host range, and *D. dorsalis* is much more widespread on the island (Keiser, Steiner, and Komasaki 1974). At higher elevations the Medfly remains dominant on peaches and other fruits, as well as on coffee, which is rarely attacked by the oriental fruit fly. Interspecific larval competition in guava fruits (main host in lowlands) favors the oriental fruit fly, which rejects interference during oviposition. Probable competition for adult food sources has also been suggested by Nishida, Harris, and Vargas (1980).

Several mortality factors have been reported to affect different developmental stages of the oriental fruit fly (Bess and Haramoto 1961; Newell and Haramoto 1968; Syed, Ghani, and Martaza 1970). Generally, in fruit flies a considerable proportion of mortality occurs during life stages found in the soil (late larval instars and pupa) (Bateman 1972). Newell and Haramoto (1968) have described predation by ants, microorganisms, excessive moisture or dryness, as well as extreme temperatures as the principal factors causing mortality or poor development of soil-inhabiting stages of *D. dorsalis* in Hawaii. According to these authors, fungi of the genus *Mucor* constitute one of the most important biotic factors in the natural control of *D. dorsalis;* although they are not pathogenic to the insect, the decomposition they cause to the fruit creates an extremely unfavorable environment that results in high mortality or otherwise reduced size of larvae and adults.

Intraspecific larval competition is another important factor in the regulation of populations of *D. dorsalis*. Once the eggs are laid, the fruit is the limiting habitat as the food source available for larval development and cycle completion (ibid). This factor, together with parasitization by introduced parasitoids were the major causes of the sharp population decline of the oriental fruit fly observed in Hawaii by 1949.

Control

The enormous threat that the oriental fruit fly posed to agriculture on the U.S. mainland prompted the development of a multidisciplinary control strategy. The research effort in the Hawaii Project was divided into five separate lines: biological control, chemical control, area control (containment and eradication), basic biology and ecology, and commodity treatment. Each line was chosen because of its unique coverage, yet the combined effort comprised a package resembling a modern integrated approach.

Based on the previous partial success of the biological control of the melon-fly and the Mediterranean fruit fly, a similar approach was attempted for the oriental fruit fly, soon after it was found in Hawaii (Clausen, Clancy, and Chock 1965). For this purpose, an extensive search for natural enemies of fruit flies was started in 1947 in the Philippines, then in the Indo-Malayan region and South America, resulting in the introduction and release of thirty-two species and varieties of natural enemies (thirty parasitic Hymenoptera and two staphylinid predators) (Bess, van den Bosch, and Haramoto 1961; Clausen, Clancy, and Chock 1965). From the fourteen taxa that were recovered from the field in Hawaii, only *Opius vandenboschi* Fullaway, *Opius oophilus* Fullaway, and *Opius longicaudatus* (Ashmead) var. *malaiaensis* Fullaway (Hymenoptera: Braconidae) were found in large numbers, parasitizing both *D. dorsalis* and *C. capitata* (Bess, van den Bosch, and Haramoto 1961). Haramoto and Bess (1970), based on Fisher's revision (1966), refer to *O. vandenboschi* as a synonym of *O. persulcatus* (Silvestri).

Opius longicaudatus and *O. vandenboschi* were released and established in 1948, followed by *O. oophilus* in 1949 (Clausen, Clancy, and Chock 1965). In the same sequence, each new establishing species displaced the other(s), leaving *O. oophilus* as the clearly dominant species by late 1950, after which the degree of parasitization leveled off at about 75 percent (van den Bosch and Haramoto 1953). The characteristic host-parasite relationships, as well as interspecific interactions among the parasites, seem to have been the causes of this phenomenon. *O. oophilus* is an egg-larval parasite; therefore it can profit entirely from any new host generation (ibid). *O. vandenboschi* and *O. longicaudatus* oviposit in the first and second to third instar larvae, respectively, therefore having a progressively more limited availability of host (ibid). Moreover, the first larva of *O. oophilus* or *O. vandenboschi* hatched in the host inhibits the development of any subsequently laid individual of the same or different species, apparently as a result of physiological reactions (van den Bosch and Haramoto 1953). Besides the effect of parasitism, *O. oophilus* also causes an appreciable mortality to the host eggs through the introduction of microorganisms, as well as through traumatic injuries during oviposition (Newell and Haramoto 1968).

As a result of this biological control program, a significant reduction of fly population and fruit infestation occurred after the peak of 1948–49 (Bess and Haramoto 1958; Haramoto and Bess 1970; Clausen 1977). The degree of infestation of both *D. dorsalis* and *C. capitata* was less than that caused by the latter alone before the introduction of the former and its parasites (Bess 1953). This was despite the fact that *Opius tryoni* (Cameron) and *Opius fullawai* (Silvestri) have become scarce as parasites of *C. capitata* since the establishment of *O. oophilus* (Bess, van den Bosch, and Haramoto 1961), apparently displacing them (Clausen, Clancy, and Chock 1965). *D. dorsalis* is, however, still considered a problem on guava (Clausen 1977).

Among the other significant contributions of the Hawaii Project was the development of three novel and complementary eradication strategies designed to eliminate incipient or low-level fruit fly populations. The development of these strategies provided tools for later fruit fly eradication projects that were successful. These tactics have also been adapted for many pest species and form the cornerstone of most modern eradication programs (Steiner 1969b).

The first of these procedures is male annihilation, which involves the use of an attractant to lure adult male flies to poisoned bait (Christenson 1963). The key ingredient is methyl eugenol, an essential plant oil, which is highly specific in attracting *Dacus* spp. (Howlett 1915). This degree of specificity suggests a coevolved relationship; however, the nature and purpose of this relationship remains obscure (Metcalf 1979).

Natural selection aside, this compound has been a powerful tool in fruit fly eradication efforts. The first bait spray ever tested in the field was methyl eugenol (Steiner 1952a). Steiner (1969b) showed that a single trap baited with methyl eugenol was as effective in catching oriental fruit flies as twenty traps baited with trimedlure for catching Mediterranean fruit flies in a one-square-kilometer area. This eventually led to the first successful eradication of a long-standing fruit fly population (Steiner et al. 1965), and methyl eugenol became a central component of most oriental fruit fly eradication projects.

As a poisoned lure, methyl eugenol was typically combined with either parathion or malathion, but by the early 1960s naled (Dibrom), became the toxicant of choice (by the late 1970s malathion had once again replaced naled in California because of concerns related to human health). The original problem with the formulation was the lack of persistance of the attractant. Early work by Steiner (1952a, 1955) showed that when flies did not succeed in consuming all the bait in a relatively short time, the remaining attractant dissipated within several days. Several formulations and trap designs were tried before satisfactory procedures were established.

The initial trap design was a feeding station that sheltered the bait from the weather. The station would quickly fill to capacity with dead flies, thus limiting access to the remaining bait (Steiner 1952a). Steiner and Lee (1955) later found that a 10 inch by 10 inch by 3/4 inch cane-fiber insulation board impregnated with an aqueous bait solution worked as well as the feeding stations and was much easier to handle. Steiner (1957b) also developed an inexpensive dry plastic trap for monitoring fruit fly populations.

Later improvements in bait formulation included the addition of glycerides of lard (monoesters of glycerides) as a thickening agent to improve adhesion and to control the amount of material delivered to a substrate (Hart et al. 1966). Cunningham, Steiner, and Ohinata (1972) showed that formulations containing thickeners were more efficacious during the first two weeks of field use compared to the standard liquid impregnated fiberboards. Ohinata, Steiner, and Cunningham (1971) developed a thixotropic formulation that was easier to prepare and apply than the thickened mixtures. Thixotropic materials can be easily applied either as a liquid from the ground in a controlled manner onto fenceposts, telephone poles, and trees or in droplet form with less selective delivery from the air. This formulation readily adheres to most surfaces while wet, then dries without losing its attractiveness. Its use tends to make traps and bait stations obsolete.

Of the eradication strategies, male annihilation is logistically the simplest to implement and economically the least expensive; but it is most effective when target populations are isolated and relatively small.

Protein-bait sprays were developed concurrently with the male annihilation procedure. These sprays were a mixture of yeast-proteins baited with a toxicant that attracted both male and female adult oriental fruit flies. Baiting with chlorinated hydrocarbons proved to be inadequate because of the slow nature of the toxic reaction (Steiner 1955), and relying on insecticides alone was clearly ineffective as a control tactic (Steiner 1957a). Bait sprays needed fast-acting insecticides, which were provided with the development of organophosphates.

Hagen and Finney (1950) showed that certain enzymatic yeast-protein hydrolysates contain amino acids necessary for fecundity. Field tests confirmed that acid and enzymatically hydrolyzed yeast-proteins were superior to partially hydrolyzed yeast-protein for attracting oriental fruit flies (Steiner 1955). A number of combinations of yeast-proteins and toxicant were tested for efficacy before hydrolyzed yeast-protein mixed with malathion was chosen as being superior (Steiner 1952b, 1955, 1957a; Harris et al. 1971). The only drawback was the discovery in urban areas that the bait formula would damage the lacquer finish on automobiles if not washed off within a day or two (Ohinata and Steiner 1967).

When properly formulated, a single pound of hydrolyzed protein can kill more than one million flies. The preferred formulation is applied as a concentrate in the form of coarse droplets. This allows the bait to compete with such attractive natural food sources as honeydew, and it remains attractive longer (Steiner 1969b). Natural enemies are spared the nontarget kill usually associated with straight insecticide spraying when exposed to protein-bait sprays (Chambers et al. 1974), and the treated environment is exposed to a small dosage of toxicant instead of a blanket of potentially disruptive material (Steiner 1955).

Protein-bait sprays cannot eradicate oriental fruit fly populations by themselves (Steiner and Lee 1955). These sprays are most effective at *controlling* fruit fly populations; consequently, it is a strategy that is best used in combination with other control procedures when eradication is the goal. The use of protein-bait sprays is moderate in cost and is no more difficult to handle than methyl eugenol sprays.

In 1938, E. F. Knipling outlined the conditions needed to eradicate an insect species given that sterile individuals in large numbers could be introduced (Steiner 1969b). Knipling's basic thesis was that overflooding a wild population with sterile adults (usually males) would lead to a decline in the population because of fewer viable matings. Eventually, eradication would follow if an adequate number of sterile individuals were released. The feasibility of the sterile male technique was demonstrated for the first time in 1954 when the screwworm (*Cochliomyia hominivorax* [Coquerel]) was eradicated from the 170 sq mi island of Curacao (Baumhover et al. 1955). The demonstration convinced individuals involved in the Hawaiian Project to seriously consider this technique as an eradication strategy for the oriental fruit fly (Steiner and Christenson 1955–56).

Effective fruit fly sterilants have included tepa, a chemosterilant (Keiser, Steiner, and Kamasaki 1965), and radiation in the form of x-rays and gamma rays. The preferred medium has been gamma radiation using a cobalt 60 source (Balock and Christenson (1955–56; Steiner 1969b). This approach requires large numbers of laboratory-reared insects that must be irradiated and then released to provide the upper overflooding ratio. According to Knipling (1955), this value was 9 : 1; it was later shown empirically to be 20 : 1 for the oriental fruit fly by Steiner (1969a); subsequent release protocol has varied from 100–200 : 1.

Mass rearing of fruit flies involves several obstacles to producing the millions of individuals needed. Rearing procedures must be efficient and economical without sacrificing the quality of the stock produced. Careful attention must be given to insure that the irradiated insects are sterile yet viable and competitive. Much research in this area has been directed toward fruit flies in general. The work specifically aimed at the oriental fruit fly includes the development of low-cost rearing medium (Tanaka et al. 1969), an evaluation of mutilation and survivorship of sterile flies in drop boxes (Harris et al. 1968), a comparison of life-table features of sterile versus nonsterile flies under a variety of conditions (Keiser and Schneider 1969a, 1969b), and determining the flight performance of irradiated flies (Sharp and Chambers 1976).

The sterile male technique is the most costly and logistically the most difficult to implement of the eradication strategies outlined. A large facility, expensive technology, special delivery systems, plus a well trained and coordinated staff with a property equipped laboratory, are all necessary.

History of Eradication Efforts

Hawaii, 1946

The discovery of the oriental fruit fly in the Hawaiian Islands launched one of the largest recorded research projects directed at an introduced insect pest (Armitage 1949; Carter 1950, 1952a, 1952b; CJLCALP 1949, 1951, 1953). The discovery of an oriental fruit fly infestation in 1946 prompted the Territorial Board of Agriculture and Forestry to initiate a search for natural enemies beginning the following year (Carter 1952b). Within two years, parasites collected from throughout the Pacific region, including the Philippines, Malaysia, Formosa (Taiwan), Thailand, India, and south and central Africa, had been introduced (CJLCALP 1949). Forty parasitoid species were eventually introduced by 1950, and four braconids of the genus *Opius* showed considerable promise as biological control agents (van den Bosch et al. 1951). It became clear by 1951 that the introduced natural enemies were controlling the oriental fruit fly populations (Carter 1952a). Biological control provided great relief to the territorial growers and reduced the risk of a mainland invasion.

The oriental fruit fly is still present in Hawaii, but it no longer represents a major pest risk to local agriculture. The huge fruit fly habitat—13,000 sq km of coastline—(Steiner 1969b) and limited resources currently make eradication an impossible task. Faint whispers of a triple fruit fly eradication attempt on all the islands are, however, occasionally heard (Nishida, Harris, and Vargas 1980).

The Hawaii Project was more than a program designed to control a local pest. Much of the research was done for the benefit of agricultural areas that had never seen an oriental fruit fly. In many ways, this project was a remote proving ground for the development of new control and eradication strategies. The knowledge and experience gained from this project proved to be valuable in many later eradication projects.

Marianas and Other Pacific Islands, 1960–1980:

The research conducted during the Hawaii Project suggested that colonies from recent invasions and small oriental fruit fly populations could be eradicated. The project provided the research background and technical expertise needed to make a first attempt at eradicating an infestation. By the late 1950s a package of eradication strategies had been developed, and the time was right for putting the theory into practice.

The Marianas Islands, a group of islands southeast of Japan and west of Hawaii, became the target for the initial eradication effort. These islands are small isolated areas containing few humans, but an annoying oriental fruit fly problem existed there. Beginning with Rota Island in 1960, a series of eradication attempts was made over a five-year period, and there were both successes and failures.

Rota Island was an ideal first candidate for eradication. It is a small island, only 33 sq mi, located within the limits of practical logistic support (thirty-seven miles north of Guam, the principal staging area, and seven and a half hours by air from the sterile fruit fly source in Hawaii). Sterile males were used in this first attempt, which targeted both the oriental and melon fruit flies. A total of 560 million oriental fruit flies were released in quantities of 6 million flies each week for two months (Steiner, Mitchell, and Baumhover 1962). The intent was to overflood the wild population with a ratio of five sterile males to each wild female; but by the end of 1961 the effort had clearly failed.

Steiner, Mitchell, and Baumhover (1962) cited the following reasons for the failure: (1) stretching scarce resources beyond their capacity—the laboratory personnel were not able to produce adequate sterile quantities of both fruit fly species simultaneously; (2) losses resulting from mechanical injuries incurred during releases; and (3) inadequate release effort as measured by the low number of recaptures (about 1 percent of all releases).

Male annihilation was implemented in 1962, and this time the effort was successful. Forty cane-fiber squares impregnated with an aqueous solution of methyl eugenol and naled were dropped in every square mile biweekly fifteen times (Steiner et al. 1965). By the end of the first generation more than 99 percent of all males found were dead; a subsequent increase in the number of females found dead soon followed (ibid). The project achieved eradication in six months and became the first documented case of a long-standing fruit fly infestation (more than thirty years) to be eliminated (ibid; Mitchell 1980).

The success at Rota Island spawned a more ambitious undertaking. This time Guam, an island ten times the size of Rota, was the target for eradication. The chance for success of the project was dramatically assisted when two huge typhoons swept the island and reduced the oriental fruit fly population by 99.3 percent (Steiner 1969b). Sterile flies were released on the ground each week at an overflooding ratio of 130 : 1. Oriental fruit flies totaling 40 million (20 million for the initial effort and 20 million to clean up reinfestations) were released for four months. Five reinfestations during the ensuing fourteen months were quickly located and treated with a combination of sterile flies, protein-bait sprays and methyl eugenol traps (Steiner et al. 1970). This project represented the second successful fruit fly eradication using the sterile male technique; the first had been against the melon fly on Rota Island in 1961 (Steiner 1969b).

Other projects in the Marianas included the islands of Saipan, Tinian, and Agiguan during 1964–65. These islands were treated with aerial-released sterile flies at an overflooding ratio of 20 : 1 for eleven and a half months without success (Steiner et al. 1970). Part of the problem included overheating of the irradiated pupae, little or no honeydew-producing insects, poor aerial dispersal of the sterile flies, and excessive predation on the released individuals (ibid). An interesting occurrence at that time was the behavior of the major predator—a toad of the genus *Bufo*—that congregated in densities of up to forty-four individuals at a time around the drop boxes and consumed up to twenty-four hundred flies each per day (ibid). Eradication was finally reached by employing the male annihilation strategy.

Some of the eradication efforts became continuing projects because of the lack of success. Male annihilation never succeeded in eliminating the oriental fruit fly from Chich, Jima, and Bonin, a group of islands south of

Japan (Christenson 1963; Habu, Iga, and Numazawa 1980). Failure of an early effort was blamed on frequent interruptions in the aerial release schedule and the ability of the fly to travel long distances between islands (Christenson 1963). In a later attempt, Habu, Iga, and Numazawa (1980) experimentally demonstrate the presence of a strain of the oriental fruit fly insensitive to methyl eugenol. Some progress was eventually made in the area using the sterile male technique; this success is, however, currently limited to the Haha Jima group of the Bonin (Ogasawa) islands (ibid).

The Japanese have been involved in a major campaign to thwart any invasion of the main island by monitoring the chain of islands to the south, which currently bridge an infested source area. The oriental fruit fly has been moving northeast from Taiwan since the early part of this century (Koyama 1980). The Japanese have established an oriental fruit fly barrier south of the island of Kyushu, where infestations have been systematically eliminated island by island (Habu, Iga, and Numazawa 1980; Koyama 1980; Moore 1973; Tanaka 1980). One of the early efforts in this campaign led to only temporary success (Moore 1973), which raised some speculation about another methyl-eugenol-insensitive strain (Ito and Iwahashi 1974). Tanaka (1980) examined this possibility and found no difference in the sensitivity of a laboratory colony of once-local flies compared to the reinfested flies.

Mainland United States, 1950–1980

The economy of Hawaii since World War II has been heavily dependent on tourism and trade with the U.S. mainland. This dependence virtually assured periodic introduction of the fly to the mainland; hence the early interest by mainland agricultural officials to stop the pest at its source. By 1949 this fly was found in produce shipments destined for California; some of the oriental fruit flies were found in "bon voyage baskets," which ironically included fruit from California, purchased by returning passengers (CJLCALP 1949). Quarantine procedures still have not restricted mainland access as intended. Hundreds of recoveries have been made since 1949; however, relatively few invasions have led to infestations requiring eradication treatment. Between 1960 and 1973 only seven infestations (two in Florida and five in California) were treated at a total cost of $80,000 (Chambers et al. 1974).

Standard detection and treatment procedures have been established as a consequence of the Hawaii Project. Susceptible agricultural areas are monitored with a methyl eugenol trap network such as that established in California in 1950 (Steiner 1952a). A positive find prompts agricultural officials to isolate and size the infestation by increasing the local trap density. Treatment consists of ground application of methyl eugenol and protein-bait sprays laced with malathion, plus soil drenches in the event larvae are present (ibid).

A major test of this protocol occurred in California during 1974–75 (CDFA 1974, 1976). A single adult fly was found in San Diego County on September 6. Eight days later larvae were found, thus representing the first confirmed larval infestation of the mainland. Consequently, several foreign countries declared a quarantine on all California produce. The presence of larvae imposed a sense of urgency not previously felt. The California Department of Food and Agriculture in cooperation with the United States Department of Agriculture took immediate steps to initiate the established treatment procedures.

Eradication treatments were started within two weeks of the initial discovery, and within another two weeks the infestation was isolated, allowing a 90-sq mi quarantine area to be established. Several new adult finds outside the original quarantine area caused the boundaries to be expanded to 315 sq mi six and a half weeks later. All infestation sites were treated with methyl eugenol and protein-bait sprays to eliminate the adults, while larvae were treated with a soil drench.

Male annihilation was terminated in April, and protein-bait spraying was terminated in May, the same month that eradication was declared—nine months after the initial discovery (Philips 1976). A total of 534 adult flies were captured, and 33 larval sites were isolated. The total cost of eradication was $750,000, which was evenly split between the state and federal governments.

Evaluation

The oriental fruit fly invasion of Hawaii represents a turning point in the short history of insect eradication. The increase in worldwide travel and trade had forced agricultural officials to keep a closer watch on important international pests while making preparations for the inevitable invasions. The discovery of the oriental fruit fly in the Hawaiian Islands stirred responsible individuals into launching an all-out attack on this new pest because it was considered a threat to new and expanding agricultural markets.

The Hawaii Project provided important basic biological information, as well as experience in biological control, and was a catalyst for many other projects. Although this project was unsuccessful in eliminating the oriental fruit fly from the islands, it did lead to the development of eradication strategies that were most effective when used in combination against small or incipient populations. The application of these strategies has gone well beyond fruit fly eradication. These techniques have been put to the test many times since their development in the 1950s. Success rates have varied, at least initially. But the observable trend in these projects seems to indicate that eradication will eventually occur if the effort is persistent.

A potentially important factor has been the discovery of strains of oriental fruit fly insensitive to methyl eugenol. This indicates that excessive and indiscriminate use of the male annihilation technique could severely reduce the effectiveness of a powerful tool and the flexibility of future eradication strategies.

The biological control program that started in 1947 resulted in a considerable lowering of populations of the oriental fruit fly, as well as the Medfly. Although the pest problems were not completely solved, the population reductions and consequently reduced fruit infestations have been significant. Future research should include the development of a conceptual model of the interactions of the oriental fruit fly and its environment. Such an approach would facilitate the understanding of biological and ecological interactions and elucidate weak and exploitable links. The model should also include pest management components so that the merits of the available eradication strategies can be assessed, given economic constraints (lack of sufficient resources) and/or biological barriers (selection for methyl eugenol insensitivity).

Although there are few favorable areas for this tropical fruit fly in the continental United States, some of them, such as southern California, represent constant possible ports of entry and bridges for further spread into the more favorable climatic areas in Mexico and Central America. Undoubtedly an infestation of those areas would be extremely difficult to eradicate. Besides imposing a new burden on the economies of the countries invaded, there would be a constant risk of the oriental fruit fly being reintroduced into the United States. Should this happen, the Hawaiian Islands experience with biological control would be very useful, especially regarding the selection of procedures to provide beneficial results.

Literature Cited

Armitage, H. M. 1949. The oriental fruit fly from the mainland viewpoint. *J. Econ. Entomol.* 42:713–16.

Balock, J. W., and L. D. Christenson. 1955–56. Effect of gamma rays from cobalt 60 on immature stages of the oriental fruit fly (*Dacus dorsalis* Hendel) and possible application to commodity treatment problems. In *Proc. Hawaiian Acad. Sci,* thirty-first ann. meeting, 8.

Bateman, M. A. 1972. Ecology of fruit flies. *Ann. Rev. Entomol.* 17:493–518.

Baumhover, A. H., A. J. Graham, B. A. Bitter, D. E. Hopkins, W. D. New, F. H. Dudley, and R. C. Bushland. 1955. Screwworm control through release of sterilized flies. *J. Econ. Entomol.* 48:462–66.

Bess, H. A. 1953. Status of *Ceratitis capitata* in Hawaii following the introduction of *Dacus dorsalis* and its parasites. *Hawaii Entomol. Soc. Prod.* 15:221–35.

Bess, H. A., and F. H. Haramoto. 1958. Biological control of the oriental fruit fly in Hawaii. *Tenth Internat. Cong. Entomol. Proc. (1956)* 4:835–40.

———. 1961. Contributions to the biology and ecology of the oriental fruit fly, *Dacus dorsalis* Hendel (Diptera: Tephritidae), in Hawaii. *Hawaii Agric. Exp. Stat. Tech. Bull.* 44.

Bess, H. A., R. van den Bosch, and F. H. Haramoto. 1961. Fruit fly parasites and their activities in Hawaii. *Hawaii Ent. Soc. Proc.* 17:367–68.

California Department of Food and Agriculture (CDFA). 1974. *Oriental fruit fly annual report.* Sacramento: CDFA.

———. 1976, 1974, 1975. *Oriental fruit fly eradication program in San Diego: A Chronology.* Sacramento: CDFA.

California Joint Legislative Committee on Agriculture and Livestock Problems (CJLCALP). 1949. *Special report on the oriental fruit fly in the Hawaii Islands,* 1–57. Sacramento: Assembly of the State of California.

———. 1951. *Special report on the current status of the oriental fruit fly in the Hawaiian islands.* Sacramento: Assembly of the State of California.

———. 1953. *Third special report on the control of the oriental fruit fly* (Dacus dorsalis) *in the Hawaiian islands.* Sacramento: Senate of the State of California.

Carter, W. 1950. The oriental fruit fly: Progress on research. *J. Econ. Entomol.* 43:677–83.

———. 1952a. Recent developments in oriental fruit fly research. *J. Econ. Entomol.* 45:274–79.

———. 1952b. Insects on fruit. *Yearbook of Agriculture.* 1952. 551–59.

Chambers, D. L., R. T. Cunningham, R. W. Lichty, and R. B. Thrailkill. 1974. Pest control by attractants: A case study demonstrating economy, specificity, and environmental acceptability. *Bioscience* 24:150–52.

Christenson, L. D. 1963. The male annihilation technique in the control of fruit flies. *Adv. Chem.,* ser. 41, 31–35.

Christenson, L. D., and R. H. Foote. 1960. Biology of fruit flies. *Ann. Rev. Entomol.* 5:171–92.

Clausen, C. P. 1977. Tephritidae (Trypedtidae, Trupaneidae): Fruit flies. In *Introduced parasites and predators of arthropod pests and weeds: A world review,* ed. C. P. Clausen, 320–35. *U.S. Dept. Agric. Handbk* 480.

Clausen, C. P., D. W. Clancy, and Q. C. Chock. 1965. Biological control of the oriental fruit fly (*Dacus dorsalis* Hendel) and other fruit flies in Hawaii. *U.S. Dept. Agric. Tech. Bull.* 1322.

Cunningham, R. T., L. F. Steiner, and K. Ohinata. 1972. Field tests of thickened sprays of methyl eugenol potentially in male-annihilation programs against oriental fruit flies. *J. Econ. Entomol.* 65:556–59.

Fisher, M. 1966. *Revision der Indo-australischen Opinnae (Hymenoptera, Braconidae).* DA. W. Junk: Den Haag.

Habu, N., M. Iga, and K. Numazawa. 1980. Progress of the eradication program of the oriental fruit fly, *Dacus dorsalis* Hendel, on the Ogasawara (Bonin) Islands. In *Proc. symp. on fruit fly problems,* Kyoto and Naha, August 9–12, Nat. Inst. Agric. Sci., Yatabe, Ibarake 305, Japan, 123–131.

Hagen, K. S. 1958. Honeydew as an adult fruit fly diet affecting reproduction. *Tenth International Cong. Entomol. Proc.* 3:25–30.

Hagen, K. S., and G. L. Finney. 1950. A food supplement for effectively increasing the fecundity of certain tephritid species. *J. Econ. Entomol.* 43:735.

Haramoto, F. H., and H. A. Bess. 1970. Recent studies on the abundance of the oriental and Mediterranean fruit flies and the status of their parasites. *Proc. Hawaiian Entomol. Soc.* 20:551–66.

Hardy, D. E. 1969. Taxonomy and distribution of the oriental fruit fly and related species (Tephritidae). *Dipteral. Proc. Hawaiian Entomol. Soc.* 20:395–428.

Harris, E. J., D. L. Chambers, L. F. Steiner, D. C. Kamakahi, and M. Komura. 1971. Mortality of tephritids attracted to guava foliage treated with either malathion or naled plus protein-hydrolysate bait. *J. Econ. Entomol.* 64:1213–16.

Harris, E. J., W. C. Mitchell, A. H. Baumhover, and L. F. Steiner. 1968. Mutilation and survival of sterile oriental fruit flies and melon flies emerging in drop boxes. *J. Econ. Entomol.* 61:493–96.

Hart, W. G., L. F. Steiner, R. T. Cunningham, S. Nakagawa, and G. Faris. 1966. Glycerides of lard as an extender for curelure, medlure, and methyl eugenol in formulation for programs of male annihilation. *J. Econ. Entomol.* 59:1395–400.

Howlett, F. M. 1915. Chemical reactions of fruit flies. *Bull. Entomol. Res.* 6:297–305.

International Institute of Entomology. 1986. *Distribution maps of pests,* Map 109.

Ito, Y., and O. Iwahashi. 1974. Ecological problems associated with an attempt to eradicate *Dacus dorsalis* (Tephritidae: Diptera) from the southern islands of Japan with a recommendation of the use of the sterile-male technique. In *The sterile-insect technique and its field applications,* Panel Proc. Ser., Internat. Atomic Energy, Vienna.

Iwahashi, O. 1972. Movement of the oriental fruit fly adults among islets of the Ogasawatan Islands. *Environ. Entomol.* 1:176–79.

Keiser, I., R. M. Kobayashi, D. H. Miyashita, E. J. Harms, E. L. Schneider, and D. L. Chambers. 1974. Suppression of Mediterranean fruit flies by oriental fruit flies in mixed infestations in guava. *J. Econ. Entomol.* 67:355–60.

Keiser, I., and E. L. Schneider. 1969a. Need for immediate sugar and ability to withstand thirst by newly emerged oriental fruit flies, and Mediterranean fruit flies untreated or sexually sterilized with gamma radiation. *J. Econ. Entomol.* 62:539–40.

———. 1969b. Longevity, resistance to deprivation of food and water, and susceptibility to malathion and DDT of oriental fruit flies, melon flies, and Mediterranean fruit flies sexually sterilized with tepa or radiation. *J. Econ. Entomol.* 62:663–67.

Keiser, I., L. F. Steiner, and H. Kamasaki. 1965. Effect of chemosterilants against the oriental fruit fly, melon fly, and Mediterranean fruit fly. *J. Econ. Entomol.* 58:682–85.

Knipling, E. F. 1955. Possibilities of insect control or eradication through the use of sexually sterile males. *J. Econ. Entomol.* 48:459–62.

Koyama, J. 1980. The Okinawa project of eradicating fruit flies. In *Proc. symp. on fruit fly problems,* Kyoto and Naha, August 9–12, Nat. Inst. Agric. Sci., Yatabe, Ibaraki 305, Japan, 99–106.

Messenger, P. S. 1960. California climatic factors and the oriental fruit fly. *California Dept. Agric. Bull.* 49:235–41.

Messenger, P. S., and N. E. Flitters. 1954. Bioclimatic studies of three species of fruit flies in Hawaii. *J. Econ. Entomol.* 47:756–65.

Metcalf, R. L. 1979. Plants, chemicals, and insects: Some aspects of coevolution. *Bull. Entomol. Soc. Am.* 25:30–35.

Mitchell, W. C. 1980. Verification of the absence of oriental fruit and melon fruit fly following an eradication program in the Mariana Islands. *Proc. Hawaiian Entomol. Soc.* 23:239–43.

Moore, I. 1973. The role of the sterile male technique in integrated control. *Bull. Org. Europ. et Medit. pour la protection des plantes* 3(3):77–83.

Newell, I. M., and F. H. Haramoto. 1968. Biotic factors influencing populations of *Dacus dorsalis* in Hawaii. *Proc. Hawaiian Entomol. Soc.* 20:81–139.

Nishida, T. E., J. Harris, and R. I. Vargas. 1980. Life systems components in relation to fruit fly eradication strategies in Hawaii. In *Proc. Symp. on fruit fly problems,* Kyoto and Naha, August 9–12, Nat. Inst. Agric. Sci., Yatabe, Ibaraki 305, Japan, 133–42.

Ohinata, K., and L. F. Steiner. 1967. Comparative damage to automobiles finishes of promising bait-spray toxicants for fruit flies. *J. Econ. Entomol.* 60:704–7.

Ohinata, K., L. F. Steiner, and R. T. Cunningham. 1971. Thixcin as an extender of poisoned male lures used to control fruit flies in Hawaii. *J. Econ. Entomol.* 64:1250–52.

Philips, F. M. 1976. Oriental fruit fly eradication, San Diego, California. *Cooper, Plant Pest Report,* Vol. 1, April 9, 141–42.

Sharp, J. L., and D. L. Chambers. 1976. Gamma irradiation effect on the flight mill performance of *Dacus dorsalis* and *Ceratitis capitata. Proc. Hawaiian Entomol. Soc.* 22:335–44.

Steiner, L. F. 1952a. Methyl eugenol as an attractant for oriental fruit fly. *J. Econ. Entomol.* 45:241–48.

———. 1952b. Fruit fly control in Hawaii with poison-bait sprays containing protein hydrolysates. *J. Econ. Entomol.* 45:838–43.

———. 1955. Bait sprays for fruit fly control. *Agric. Chem.* 10:113–15.

———. 1957a. Field evaluation of oriental fruit fly insecticides in Hawaii. *J. Econ. Entomol.* 50:16–24.

———. 1957b. Low-cost plastic fruit fly trap. *J. Econ. Entomol.* 50:508–9.

———. 1969a. A method of estimating the size of native populations of oriental, melon, and Mediterranean fruit flies, to establish the overflooding ratios required for sterile-male releases. *J. Econ. Entomol.* 62:4–7.

———. 1969b. Control and eradication of fruit flies on citrus. In *Proc. First Internat. Citrus Symp.,* Vol. 2, 881–87.

Steiner, L. F., and L. D. Christenson. 1955–56. Potential usefulness of the sterile fly release method in fruit fly eradication programs. *Proc. Hawaiian Acad. Sci.,* thirty-first annual meeting.

Steiner, L. F., W. G. Hart, E. J. Harris, R. T. Cunningham, K. Ohinata, and D. C. Kamakahi. 1970. Eradication of the oriental fruit fly from the Mariana Islands by the methods of male annihilation and sterile insect release. *J. Econ. Entomol.* 63:131–35.

Steiner, L. F., and R. K. S. Lee. 1955. Large-area tests of a male-annihilation method for Oriental fruit fly control. *J. Econ. Entomol.* 48:311–17.

Steiner, L. F., W. C. Mitchell, and A. H. Baumhover. 1962. Progress of fruit-fly control by irradiation sterilization in Hawaii and the Marianas Islands. *International Journal of Applied Radiation and Isotopes* 13:427–34.

Steiner, L. F., W. C. Mitchell, E. J. Harris, T. T. Kozuma, and M. S. Fujimoto. 1965. Oriental fruit fly eradication by annihilation. *J. Econ. Entomol.* 58:961–64.

Steiner, L. F., G. G. Rohwer, E. L. Ayers, and L. D. Christenson. 1961. The role of attractants in the recent Mediterranean fruit fly eradication program in Florida. *J. Econ. Entomol.* 54:30–35.

Syed, R. A., M. A. Ghani, and M. Martaza. 1970. Studies on the Trypetids and their natural enemies in West Pakistan IV. Further observations on *Dacus (Strumeta) dorsalis* Hendel. *Tech. Bull.* CIBC 13, 17–30.

Tanaka, A. 1980. Present status of fruit fly control in Kagoshima prefecture. In *Proc. symp. on fruit fly problems,* Kyoto and Naha, August 9–12, Nat. Inst. Agric. Sci. Yatabe, Ibaraki 305, Japan, 107–21.

Tanaka, N., L. F. Steiner, K. Ohinata, and R. Okamoto. 1969. Low-cost larval rearing medium for mass production of oriental and Mediterranean fruit flies. *J. Econ. Entomol.* 62:967–68.

van den Bosch, R., H. A. Bess, and F. H. Haramoto. 1951. Status of oriental fruit fly parasites in Hawaii. *J. Econ. Entomol.* 44:753–59.

van den Bosch, R., and F. H. Haramoto. 1953. Competition among parasites of the oriental fruit fly. *Proc. Hawaiian Entomol. Soc.* 15:201–6.

12

Citrus Canker

Laura D. Merrill

The eradication of citrus canker, *Xanthomonas campestris* pv. *citri* (Hasse) Dye, in the United States is considered an important example of a completely successful eradication program. When discovered in Florida in 1913, this disease aroused intense concern because of the severity of the damage to grapefruit, *Citrus paradisi* Macfad. (Stevens 1914a, 1914b), a species later found to be one of the more susceptible (Dopson 1964). The disease is believed to have arrived in the United States on infected nursery stock from Japan (Fawcett 1936). On the basis of evidence from herbarium specimens and distribution records, Fawcett suggested that the disease was native to India and Java, from which it spread to the Far East and later to other citrus-growing regions of the world. The present distribution of the pathogen includes parts of Asia, India, the Soviet Union, the Philippine Islands, and Latin America (Calavan 1956; Garnsey et al. 1979; Mkervali 1980; Russell 1977). The statement in the text by Walker (1969, 723) that the disease is endemic on noncommercial citrus plantings in Alabama, Mississippi, and Louisiana is probably incorrect, as other reports indicate that, prior to 1984, infected trees had not been found in the United States since 1940 (Sinclair 1968) or 1943 (Dopson 1964). The pathogen is also reported to have been eradicated from South Africa (Calavan 1956; Doidge 1930), Australia (Klotz 1973), Mozambique, and portions of New Zealand (Garnsey et al. 1979). Citrus canker has never been found in California or Arizona (Calavan 1956).

Much of the literature on citrus canker eradication is anecdotal and, perhaps because of its historical era, is not presented in the scientific format used today. Most publications supported the eradication effort, and only Loucks (1934) discussed grower opposition to the program. Newspaper files and other public forums might provide an interesting perspective on the project. These sources are not, however, discussed in this chapter.

I thank L. E. Caltagirone for his helpful criticisms and N. Davidson, A. M. Liebhold, J. R. Parmeter, S. Tait, and D. L. Wood for their careful reviews of the manuscript. I particularly thank D. Zagory for his review and for his expert advice.

Life History and Habits

The disease is caused by the bacterium *Xanthomonas campestris* pv. *citri* (Hasse 1915; Young et al. 1978). Synonyms include *Pseudomonas citri, Bacillus citri, Phytomonas citri,* and *Xanthomonas citri.* This bacterium is a gram negative rod with a single polar flagellum (Calavan 1956). Like most plant pathogenic bacteria, it requires wounds or natural openings (stomata or lenticels) for entry into the host (Agrios 1969, 331). The pathogen overwinters in lesions on infected fruit, leaves, twigs, and other plant parts (Calavan 1956; Koizumi 1979). In Florida, the pathogen was apparently unable to survive saprophytically and quickly disappeared from orchard soils (Fawcett 1936). Recent reports indicate that this bacterium may survive for limited periods on roots of grass, in weeds, in soil, and in dead plant material (Goto, Ohta, and Okabe 1975a, 1975b; Pereira et al. 1976, 1978; Stall et al. 1981). Experiments in South Africa indicated that the pathogen could survive saprophytically in the soil for several years (Doidge 1929), but this has not been confirmed.

Excluding the above reports of survival in grass and weeds, the only known hosts of the pathogen are *Citrus* and such close relatives as trifoliate orange (D. Zagory, personal communication, 1984). Cultivars vary in susceptibility. Grapefruit, trifoliate orange (*Citrus* sp.), sweet orange (*Citrus sinensis* [L.]), and lemon (*Citrus limon* [L.]) are highly susceptible; mandarin orange (*Citrus reticulata* Blanco) is moderately resistant, and kumquat (*Fortunella* sp.) is highly resistant (Fawcett 1936; Garnsey et al. 1979). A cultivar may vary in its susceptibility to different strains of the pathogen (Stall and Seymour 1983). Calavan (1956) listed other susceptible genera and stated his belief that other, untested North American species are probably also susceptible. These species apparently did not become infected during the thirty years that the pathogen was present in the United States.

Damage and Epidemiology

X. campestris pv. *citri* causes lesions on leaves, twigs, and fruits of susceptible plants. Lesions begin as white, pinhead sized eruptions, change to tan, then become brown with a yellow halolike margin, and attain a size of 3–5 mm on leaves and up to 15 mm on fruit (Klotz 1973; Stall and Seymour 1983; D. Zagory, personal communication, 1984). Cankers on the leaves may result in defoliation, probably as a result of bacterial ethylene production (Goto and Yaguchi 1979; Goto, Yaguchi, and Hyodo 1980). Cankers on the fruit may render the crop unmarketable for cosmetic reasons (Walker 1969) and may serve as entry points for secondary rot fungi (Stevens 1914b). The disease may also cause dieback and premature fruit drop (Garnsey et al. 1979). In many countries canker is considered to be the most destructive disease of citrus (Calavan 1956).

Two aspects of the ecology of the citrus canker organism are of particular importance with respect to its eradication. First, the pathogen usually requires moisture both for survival outside the plant and for infection. Because of this, many workers believe that it could not survive in the commercial citrus growing regions of California, where there is little or no precipitation during the warm months in which the pathogen would be active (Fawcett 1936; Klotz 1973; Peltier and Frederich 1926), although sprinklers might provide the needed moisture. Also, the pathogen has been reported to survive for several months in air-dried soils (Goto, Ohta, and Okabe 1975b), on sawdust and other materials, and in sterilized distilled water (Loucks 1934). Second, like most bacterial plant pathogens and unlike many fungi, the bacterium does not have a resistant airborne spore state and can only be spread from plant to plant by animals (including humans), running water, and windblown and splashing rain. Harrison, Brewer, and Merrill (1980) discussed the special problems that bacterial pathogens encounter while moving between and infecting host plants. Humans were undoubtedly responsible for much of the spread in the United States by moving the pathogen from place to place on infected nursery stock (Stevens 1915). Recent work has shown that the pathogen can be spread on pruning shears and on the hands of fruit pickers (Messina 1977). There is evidence that the pathogen may be carried for considerable distances via running water (Doidge 1929). Peltier and Frederich (1926) listed the following prerequisites for successful infection: (1) temperatures between 20°C and 35°C (optimum at 30°C), (2) free moisture on the plant surface for twenty minutes or more, and (3) rapidly growing leaves, stems, or young fruits.

Control

Control of the disease is extremely difficult and continues to be a problem in areas of the world where the pathogen has not been eradicated, such as Asia, India, and Latin America. Control methods include sanitation, planting of resistant cultivars (Fawcett 1936; Kuhara 1978), adjusting cultural practices (for example, controlling fertilization and irrigation to reduce succulence during the warm, moist season and to increase growth during the dry season) (Fawcett 1936), planting of windbreaks to reduce disease spread and to prevent wind damage to the host (ibid; Kuhara 1978), controlling possible insect vectors (Fawcett 1936), and spraying with antibiotics (Kuhara 1978) or with copper compounds, including Bordeaux mixture (Klotz 1973; Koizumi 1979; Mkervali 1980; Stall et al. 1981). Kuhara (1978) described an integrated pest management system in use in Japan, involving resistant varieties, disease forecasting, and cultural and chemical controls. Chemicals currently in use are unsatisfactory because of toxicity (for example, Bordeaux mixture), expense, rapid breakdown (for example, streptomycin) (Kuhara 1978), and bacterial resistance. As of 1979 there were no chemicals registered in the United States for use against this bacterium, although chemicals could be used under special emergency provisions with appropriate justification (Garnsey et al. 1979). Control efforts in Argentina using tribasic copper sulfate have been promising (Miller et al. 1980; Stall, Canteros de Echenique, and Marco; Zagory et al. 1984). Other fungicides have been tested as controls for this bacterium (Beniwal and Chaubey 1976). Control measures on susceptible cultivars are not adequate to permit profitable production, and more than 99 percent of the citrus trees planted in the United States have been classified as susceptible (Stall and Seymour 1983).

History of Eradication Efforts

In the United States, citrus canker was of initial concern in Florida, where eradication was subsequently begun (Berger 1914). The bacterium is believed to have been introduced to the Gulf States in about 1910 (Dopson 1964) or perhaps several years earlier (Sinclair 1968) on Japanese nursery stock (Peltier and Frederich 1926). In Florida, infected specimens were sent to the Experiment Station in 1912, but the disease was misidentified as citrus scab, a fungal disease (Berger 1914; Stevens 1914a). The following year, citrus canker was recognized as a new disease after being found in a nursery in southern Dade County, Florida. During a March 1914 survey, the disease was also discovered in Alabama, Louisiana, Mississippi, and Texas, but not in a location visited in Mexico (Berger 1914). Quarantine and eradication were recommended (Stevens 1914b).

In spring 1914 the Florida State Board of Control adopted Rule No. 44, which prohibited the importation of citrus stock into the state, although no government funds were available for enforcement (ibid). The Florida Growers' and Shippers' League was "instrumental" in arousing grower interest in canker (Calavan 1956) and contributed $2,000 to pay the expenses of Frank Stirling, an Agricultural Extension pathologist assigned to work on the eradication program in Florida. Subsequently, in September 1914, the governor of Florida allocated $1,000 to the project (Berger 1914). It was initially thought that the disease would be found on no more than ten to twelve properties and that the eradication could be completed within two months. However, Stirling (1914) found infections on ninety-five properties in Dade County alone. He was rapidly able to enlist the aid of growers for the eradication, perhaps because 80 percent of the trees in the most seriously affected area were grapefruit, on which it was believed that all fruit would be reduced to culls by the disease.

At first, treatment consisted of cutting back and defoliating infected trees and painting their trunks with Bordeaux mixture or chlorinated anthracene oil (Carbolineum). It was soon noted that new growth on these treated trees was infected; thus it was concluded that this treatment was ineffective, and the development of new infections on other trees suggested that the eradication workers were spreading the disease (ibid). In addition, experimental work indicated that infected leaves coated with Bordeaux mixture were capable of transmitting the disease (Stevens 1914a). In summer 1914 the eradication workers began spraying a flaming mixture of kerosene and crude oil onto the infected trees, the grass around the trees, and the soil, until the trees were charred. This method avoided contaminating tools. Upon leaving a grove, workers removed their clothes and dipped them in a solution of mercuric chloride. At the height of the 1914 eradication season, fifty men, all growers, were employed on the project at a total cost of $100/day (Stirling 1914). Stirling estimated that if the eradication had been undertaken six months later, the final cost would have been more than $100,000. The eradication in Florida alone eventually cost over $2.5 million (presumably 1914–31 dollars) (Rhoads and DeBusk 1931).

In early 1915, the second year of the eradication, the United States Department of Agriculture (USDA) passed Quarantine No. 19, which prohibited the importation of all nursery stock of citrus and related plants from all parts of the world. Certain states passed similar regulations. Citrus canker was responsible for the first federal appropriation ever made for a plant disease. These funds were immediately used to start a joint federal-state eradication program in Florida. Similar programs were soon organized in other gulf states, South Carolina, and Georgia (Dopson 1964). Also in 1915, a USDA researcher, Clara Hasse, discovered that the causal organism was a bacterium, not a fungus, and named it *Pseudomonas citri* (Hasse 1915). By then it was understood that spread occurred primarily via shipment of infected stock. Eradication protocol in Florida involved burning infected trees as soon as they were discovered (Stevens 1915). Similar methods were employed in Louisiana (Sinclair 1968). If trees were small, the entire grove was burned. All properties where disease was discovered were placed under quarantine. When there was evidence of spread within the grove, the quarantine was extended to cover the area within a one-mile radius of the infection (Stevens 1915).

By the end of 1916, most of the disease in Florida had been eliminated, and workers only had to contend with occasional new infections, the last of which was found in 1927. Until 1984, no infections had been found since 1917 in South Carolina, 1918 in Georgia, 1922 in Mississippi, 1927 in Alabama,

1940 in Louisiana (Sinclair 1968), and 1940 (Sinclair 1968) or 1943 (Dopson 1964) in Texas. In 1944, the federal and state cooperative program was discontinued because of a lack of funds. This program was resumed in Texas in 1947, however, because of a California quarantine of Texas grapefruit. Surveys were conducted in Texas for five years until funding was again terminated. In this latter inspection, 4.7 million trees on 26,000 properties in 81 counties were inspected; no canker was found. The cooperative program was similarly resumed in Louisiana in 1949 to inspect abandoned and escaped trees. In three annual surveys, no canker was found on the 281,000 trees inspected on 3180 properties (Dopson 1964).

Sinclair (1968) discussed difficulties in the Louisiana eradication program, including large numbers of abandoned groves, ornamental citrus plantings, wild seedlings, swampy lands, and the lack of an extensive citrus industry to support the eradication campaign. Fortunately, the Work Projects Administration provided the unskilled labor needed for the intensive seedling searches, which required the workers to crawl across fields shoulder to shoulder on their hands and knees.

The estimated costs of the eradication program in Florida are as follows: as of 1931, $2.5 million had been spent on the eradication (including federal, state, and private funds) and the *disease* was estimated to have cost $6.5 million, including the cost of 257,745 grove trees and 3,093,110 nursery trees that were destroyed (Rhoads and DeBusk 1931). The above figures are presumably in 1914–31 dollars. Based on the wholesale price index (United States Department of Commerce 1975, 199) and the consumer price index (ibid., 210–11), where 1967 dollars are the standard (that is, all 1967 price indices equal 1000.0), the eradication program in Florida cost $3.1–8.3 million, and the disease as a whole is estimated to have cost $8.2–22 million in that state, depending on which years (1914–31) and which index are used. The current issue of *Statistical Abstract of the United States* (United States Department of Commerce 1984) can be consulted to estimate these costs in current funds, but a more accurate estimate of the expense of an eradication program today could be made from actual labor and production costs and current market value of citrus fruits.

The cost of the eradication program in all of the southeastern United States has been estimated at $5 million. Total damage from the disease is estimated to have exceeded $40 million, including the cost of the eradication program, lost production, destroyed trees, and losses resulting from quarantine restrictions (Calavan 1956). The bases for these estimates were not described. Costs of present-day quarantines for this disease have not been estimated.

In South Africa eradication also involved the destruction of infected trees and was dependent on cooperation from growers (Doidge 1929, 1930).

A twenty-year eradication campaign in southern Brazil has been only partially successful (Gonzalez 1978), possibly because of reintroductions from
neighboring countries (Rossetti 1977) or extended survival of the pathogen
on an endemic grass (Garnsey et al. 1979; Pereira et al. 1976).

At present, U.S. Quarantine No. 19 prohibits importation of citrus plant
parts (except seed), and Quarantine No. 28 prohibits fruit from infested
countries (with the exceptions noted below). The disease was intercepted
2,603 times at U.S. ports between 1973 and 1978 (Garnsey et al. 1979), which
may explain the concern of Florida growers and the California Department
of Food and Agriculture, which has placed citrus canker on its list of eight
most dangerous exotic pests (P. Stoddard, personal communication, 1983).
These canker interceptions primarily involved stores of ships and passenger
baggage (Stall and Seymour 1983). Garnsey et al. (1979) believed that the
disease would eventually be reestablished in the U.S. On August 27, 1984
the disease was discovered in a nursery in Highlands Country, Florida.
Within a month the disease was found in three other nurseries, and an
eradication program was initiated. In September 1984 it was estimated that
the disease had been present for at least a year, although the date and means
of introduction were unknown (C. Krass, personal communication, 1984).
Sun (1984) described this most recent epidemic, which involved a previously unknown strain of the canker bacterium. Subsequent work has
shown that the strain of *X. campestris* involved is much less virulent than the
strain present in 1914, and eradication of the present strain is considered
impractical and unnecessary.

Evaluation

The 1914–43 program in the United States has been termed "an outstand-
ing example of a large-scale eradication program which was entirely successful" (Walker 1969, 737). True eradication—where all population
members were destroyed throughout the south-eastern United States—
was undoubtedly accomplished. The success of this eradication can be
attributed to six major factors. First, aggressive action—almost overreac-
tion—was taken within a few years of the time the disease was thought to
have arrived in North America. Well-established pests, present for long
periods, are thought to be much more difficult or impossible to eradicate
(Newsom 1978). Second, the pathogen is not capable of moving from place
to place on its own and was dependent upon humans for most of its dissemination; knowledge of this aspect of the bacterium's ecology arrested
the spread of the disease. Third, the eradication was aided by strong grower
support and high worker motivation; at least initially, many eradication
workers were citrus growers. Fourth, the pathogen was an introduced
parasite of an introduced plant, and both the bacterium and its host were

geographically isolated from their parent populations. Fifth, the bacterium probably does not survive for long periods away from its host (except for the reports discussed earlier), and the pathogen can only breach the host surface under special environmental conditions. Sixth, the host range of the pathogen is very restricted.

Other factors may have also contributed to the success of the eradication program. The early misidentification of the casual organism as a fungus (Stevens 1914b; Stirling 1914) may have led to stronger precautions than might have been taken otherwise, as a fungus could be more easily spread via wind-borne propagules than a bacterium. Furthermore, the scope of the eradication project was not recognized at first; the gross underestimations of the cost of the eradication may have helped in promoting the project. It is also possible that the strain of bacterium present in this country was less adapted for saprophytic survival or different in some other way from strains present in parts of the world where eradication has been unsuccessful. At least three strains of this pathogen, which differ in virulence and host range, have since been identified. All three strains have been found in South America, where eradication has been unsuccessful (Civerolo and Fan 1982; Rossetti 1977). Descriptions of the disease in the United States suggest that only the common Asian strain (cancrosis A), generally thought to be the most virulent (Garnsey et al. 1979; Stall and Seymour 1983), was present here. Finally, such unknown factors as climate or the lack of insect vectors may have sufficiently reduced survivorship and reproductive success of the bacterium to allow its eradication.

That the eradication was successful despite requiring nearly thirty years of effort is a tribute to the persistence of the individuals and government agencies involved. The absence of support from the pesticide industry, whose products were not used in the citrus canker eradication program, precluded the type of political support seen in some insect eradication attempts.

Eradication of a pest may produce two types of negative impact on the environment. The first is the general pollution of the treated area with a toxic biocide, and the second, disruption of the community of which the pest was a member. Unlike most insect eradication programs, citrus canker eradication did not involve extensive use of pesticides. Some spills of mercuric chloride (used in sterilizing workers' clothing) probably occurred, and burning of groves undoubtedly contributed to air pollution. Even considering the medical consequences (not mentioned in the literature) for eradication workers from wearing mercury-saturated clothing and from breathing kerosene fumes and smoke, the hazards of this program seem trivial compared with the widespread contamination of human tissues and aquatic organisms from the mirex used in attempts to eradicate imported fire ants (Hinkle 1982) (see chapter 13, this volume).

Rabb (1978) discussed the impact of the removal of a species from its biological community, including effects on energy flow, on other competing pest species, on other pests sharing the same natural enemies, and on ecosystem stability. We have no information on this aspect of citrus canker eradication. Presumably, the pathogen was not well established when the eradication was initiated and thus had not developed extensive associations with other species. When the last infected trees were burned about thirty years (and thousands of bacterial generations) later, conditions may have changed greatly. Because citrus canker is a relatively host-specific parasite of an introduced plant, however, it may be specious to discuss interspecific relationships in this country. Finally, the eradication technique used was species specific. Trees supporting other organisms but not *X. campestris* pv. *citri* were not burned.

The economic impact of citrus canker eradication has never been fully assessed. A careful cost-benefit analysis, including losses mentioned by Calavan (1956), the cost of maintaining quarantines, of inspections, and of the social costs resulting from unemployment following burning of groves, would be valuable in determining the usefulness of eradication as a pest management strategy.

The long-term success of any eradication program depends upon subsequent exclusion of the pest. To this end quarantines are essential; they can, however, also be misused as economic weapons. For example, Garnsey et al. (1979) noted the concern of Florida citrus growers about exceptions to U.S. Quarantine No. 28. These exceptions permit importation of "Unshu" oranges from Japan into Alaska, Hawaii, Idaho, Montana, Oregon, and Washington, with stringent safeguards (ibid; Stall and Seymour 1983). The USDA investigated these concerns and concluded that these exceptions present a negligible risk to the Florida citrus industry (Garnsey et al. 1979). Similarly, one can speculate that the California quarantine of Texas grapefruit in 1947 was more economically than biologically motivated.

As a pest management strategy, eradication has definite emotional appeal. A success may raise false expectations among politicians and the public with regard to the control of such other pests as the imported fire ants (chapter 13, this volume) the Japanese beetle (chapter 7), and the gypsy moth (chapter 15), which realistically cannot be eradicated from the United States. Scientists have the obligation to educate the public on the differences between pests that can and cannot be eradicated and on the potential environmental impact of eradication attempts. Few of us agree with Eden (1978, 53) that it is desirable to "drop an atomic bomb . . . to end the war." A more rational approach to most pest problems is the use of integrated pest management techniques (Rabb 1978) and educated coexistence.

Literature Cited

Agrios, G. N. 1969. *Plant pathology.* Academic Press, New York.

Beniwal, S. P. S., and S. N. Chaubey. 1976. Thiram, a fungicide effective against *Xanthomonas citri* (Hasse) Dowson. *Pesticides* 10:31, 36.

Berger, E. W. 1914. Citrus canker (2): History of citrus canker. *Fla. Agr. Exp. Sta. Bull.* 124:27–30.

Calavan, E. C. 1956. Citrus canker—a bacterial disease caused by *Xanthomonas citri. Calif. Dept. Agr. Bull.* 45:259–62.

Civerolo, E. L., and F. Fan. 1982. *Xanthomonas campestris* pv. *citri* detection and identification by enzyme-linked immunosorbent assay. *Plant Disease* 66:231–36.

Doidge, E. M. 1929. Further citrus canker studies. *Dept. Agr. Union So. Africa Bull.* 51.

————. 1930. Citrus canker eradication. *Farming in So. Africa* 5:75–76.

Dopson, R. N., Jr. 1964. The eradication of citrus canker. *Plant Dis. Reptr.* 48:30–31.

Eden, W. G. 1978. Eradication of plant pests—pro. *Bull. Entomol. Soc. Amer.* 24:52–54.

Fawcett, H. S. 1936. Citrus canker. In *Citrus diseases and their Control,* 2d ed., 237–49. N.Y.: McGraw Hill.

Garnsey, S. M., E. P. DuCharme, J. W. Lightfield, C. P. Seymour, and J. T. Griffiths. 1979. Citrus canker: Preventative action to protect the U. S. citrus industry. *Citrus Ind.* 60:5–13.

Gonzalez, R. H. 1978. Introduction and spread of agricultural pests in Latin America: analysis and prospects. *Food and Agriculture Organization (FAO) Plant Protection Bull.* 26:41–52.

Goto, M., K. Ohta, and N. Okabe. 1975a. Studies on saprophytic survival of *Xanthomonas citri* (Hasse) Dowson (1): Detection of the bacterium from a grass (*Zoysia japonica*). *Ann. Phytopathol. Soc. Japan* 41:9–14.

————. 1975b. Studies on saprophytic survival of *Xanthomonas citri* (Hasse) Dowson (2): Longevity and survival density of the bacterium on artificially infested weeds, plant residues and soils. *Ann. Phytopathol. Soc. Japan* 41:141–47.

Goto, M., and Y. Yaguchi. 1979. Relationship between defoliation and disease severity in citrus canker. *Ann. Phytopathol. Soc. Japan* 45:689–94.

Goto, M., Y. Yaguchi, and H. Hyodo. 1980. Ethylene production in citrus leaves infected with *Xanthomonas citri* and its relation to defoliation. *Physiol. Plant Pathol.* 16:343–50.

Harrison, M. D., J. W. Brewer, and L. D. Merrill. 1980. Insect involvement in the transmission of bacterial pathogens. In *Vectors of plant pathogens,* ed. Harris, K. F. and K. Maramorosch, 201–92. N.Y.: Academic Press.

Hasse, C. H. 1915. *Pseudomonas citri,* the cause of citrus canker. *J. Agr. Res.* 4:97–100.

Hinkle, M. K. 1982. Impact of the imported fire ant control programs on wildlife and quality of the environment. In *Proceedings of the symposium on the imported fire ant,* June 7–10, 1982, Atlanta, Georgia, ed. S. L. Battenfield, 130–143. Washington D.C.: USDA, Animal, Plant Health Inspection Service and Environmental Protection Agency.

Klotz, L. J. 1973. *Color handbook of citrus diseases.* Berkeley: Univ. of Calif. Press.

Koizumi, M. 1979. Relation of temperature to the development of citrus canker lesions in the spring. *Proc. Internat. Soc. Citriculture* 3:924–28.

Kuhara, S. 1978. Present epidemic status and control of the citrus canker disease (*Xanthomonas citri* (Hasse) Dowson) in Japan. *Rev. Plant Prot. Res.* 11:132–42.

Loucks, K. W. 1934. Citrus canker and its eradication in Florida. Unpublished manuscript, original copy in the files of the Division of Plant Industry, Florida Department of Agriculture, Gainsville.

Messina, M. A. 1977. Serological methods employed in the study of citrus canker in Argentina (2): First experimental evidence of the means of ecological spread. *Serie Tecnica-Estacion Experimental Regional Agropecuaria Parana 48.*

Miller, J. W., R. E. Stall, G. M. Marco, and B. I. Canteros de Echenique. 1980. Informe tecnico no. 2, Instituto Nacional de Tecnilogia Agropecuaria, Bella Vista, Corrientes, Argentina.

Mkervali, V. G. 1980. Diseases of citrus Trees. *Zashchita Rastenii* 8:46–47.

Newsom, L. D. 1978. Eradication of plant pests—con. *Bull. Entomol. Soc. Amer.* 24:35–40.

Peltier, G. L., and W. J. Frederich. 1926. Effects of weather on the world distribution and prevalence of citrus canker and citrus scab. *J. Agric. Res.* 32:147–64.

Pereira, A. L. G., K. Watanabe, A. G. Zagato, and P. L. Cianciulli. 1976. Survival of *Xanthomonas citri* (Hasse) Dowson on sourgrass (*Trichachne insularis* [L.] Nees) from eradicated orchards in the State of Sao Paulo. *Biologico* 42:217–21.

———. 1978. The survival of *Xanthomonas citri* (Hasse) Dowson, the causal agent of "citris canker" in the rhizosphere of the "Guineagrass plants" (*Panicum maximum* Jacq.). *Biologico* 44:135–38.

Rabb, R. L. 1978. Eradication of plant pests—con. *Bull. Entomol. Soc. Amer.* 24:40–44.

Rhoads, A. S., and E. F. DeBusk. 1931. Diseases of citrus in Florida. *Fla. Agric. Exp. Sta. Bull.* 229.

Rossetti, V. 1977. Citrus canker in Latin America: A review. *Proc. Internat. Soc. of Citriculture* 3:918–24.

Sinclair, J. B. 1968. Eradication of citrus canker from Louisiana. *Plant Dis. Reptr.* 52:667–70.

Stall, R. E., B. I. Canteros de Echenique, and G. M. Marco. 1979. Cancrosis de los citrus. Informe tecnico no. 1, Instituto Nacional de Tecnologia Agropecuaria, Bella Vista, Corrientes, Argentina.

Stall, R. E., J. W. Miller, G. M. Marco, and B. I. Canteros de Echenique. 1981. Population dynamics of *Xanthomonas citri* causing cancrosis of citrus in Argentina. *Proc. Fla. State Hort. Soc.* 93:10–14.

Stall, R. E., and C. P. Seymour. 1983. Canker, a threat to citrus in the Gulf-Coast states. *Plant Diseases* 67:581–85.

Stevens, H. E. 1914a. Citrus canker. A preliminary bulletin. *Fla. Agric. Expt. Sta. Bull.* 122:113–18.

———. 1914b. Citrus canker (2): Studies of citrus canker. *Fla. Agric. Expt. Sta. Bull.* 124:31–43.

———. 1915. Citrus canker (3) *Fla. Agr. Expt. Sta. Bull.* 128.

Stirling, F. 1914. Citrus canker (2): Eradication of citrus canker. *Fla. Agric. Expt. Sta. Bull.* 124:44–53.

Sun, M. 1984. The mystery of Florida's citrus canker. *Science* 226:322–23.

United States Department of Commerce. 1975. Historical statistics of the United States: Colonial times to 1970. Washington, D. C.: Bureau of the Census.

———. 1984. Purchasing Power of the Dollar, 1940–1983. In *Statistical abstract of the United States,* 104th ed., 484. Washington, D. C.: Bureau of the Census.

Walker, J. C. 1969. *Plant pathology.* 3d ed. N.Y.: McGraw-Hill.

Young, J. M., D. W. Dye, J. F. Bradbury, C. G. Panagopoulos, and C. F. Robbs. 1978. A proposed nomenclature and classification for plant pathogenic bacteria. *New Zealand J. Agric. Res.* 21:153–77.

Zagory, D., B. I. Canteros de Echenique, R. E. Stall, and J. W. Miller. 1984. Cancrosis de los citrus. Informe tecnico no. 3, Instituto Nacional de Tecnilogia Agropecuaria, Bella Vista, Corrientes, Argentina.

13

Imported Fire Ants

Nita A. Davidson and Nick D. Stone

The history of the imported fire ant is one of controversy. Its pest status, public health status, and the control policies taken against it have been hotly debated topics since 1957 when the United States Department of Agriculture (USDA) first decided to attempt eradication. Recently the USDA has again begun to operate a large-scale, expensive control effort against fire ants, thereby stimulating renewed interest in the ant, past control efforts, and the efficacy of the current approach.

Imported fire ants are actually two species of South American ants in the genus *Solenopsis*. *Solenopsis richteri* Forel, the less common black imported fire ant, was mysteriously introduced into the United States in 1918 through the port of Mobile, Alabama, but escaped recognition as a new arrival until 1930 (Summerlin and Green 1977). In South America, its range covers extensive areas in Brazil, Uruguay, and Argentina (Buren et al. 1974), but its spread in the United States today includes only northeastern Mississippi and northwestern Alabama. This limited range is probably due to competition from its relative, *Solenopsis invicta* Buren. The two fire ant species, *Solenopsis invicta* and *S. richteri*, inhabit approximately 93,120,000 hectares in nine southern states (fig. 13.1), making them a familiar feature of life in these areas (Lofgren 1986). Within that range there are about 10 billion colonies (Metcalf et al. 1982).

These ants are feared because when a nest is disturbed the ants swarm over any nearby object, delivering multiple, painful stings to the intruder. This aggressiveness extends only to a small area around the nest, which is a conspicuous mound (Sterling 1978). The behavior is, in fact, not unlike that of the well-known winged relatives of the ant—bees and wasps.

The ant is ubiquitous in agriculture. It has been reported in virtually every crop grown within the infested states, and while it certainly causes

We would like to thank Gordon Frankie, Laura Merrill, and Walter Tschinkel for their careful and insightful reviews of the manuscript. Denzel Ferguson and Carolyn Carr provided valuable suggestions.

Figure 13.1 Distribution of the imported fire ant, *Solenopsis invicta* and *S. richteri,* in the United States (adapted from Lofgren 1986).

damage, it is also sometimes beneficial because it preys on a wide variety of insects.

Except for the range of the fire ant, the situation today is similar to what it was in 1957 when the USDA attacked fire ants with the largest and most devastating eradication program ever undertaken against an introduced insect pest. More than twenty-five years and close to $200 million later (Oliver, Reagan, and Burns 1979), the situation is now worse than ever.

Life History and Habits

Fire ants have a typical ant life cycle. Mating occurs during a swarming flight shortly after a warm rain. Inseminated females land, break off their wings, and search for a suitable nesting site. One to several mated queens may initiate a nest cooperatively, digging a burrow two to five inches deep that ends in a widened chamber. In the next month, the queen produces the first brood (called minims) without foraging, relying on stored nutrients and her degenerating flight muscles.

During the growth phase of the colony the number of workers increase greatly, and the mound first becomes visible. Generally, workers kill all but one of the queens during this stage. In the western reaches of their range, however, fire ant colonies are often found with hundreds of laying queens per nest (Tschinkel 1982). Polygynous colonies have also been located in northern Georgia and may be common elsewhere in the range (W. R. Tschinkel, personal communication, 1985). The growth phase lasts about two years, but reproduction can occur as few as six months after nest initiation.

The alate reproductive caste is produced throughout the year, but they leave the nest only in large coordinated swarms. Timing seems to depend on high temperature and humidity (Tschinkel 1982; Summerlin and Green 1977), often occurring a few days after a warm rain. In areas with cold winters the mating swarms become facultatively seasonal, with peak swarming between May and August. Mating occurs on the wing, and the flight generally disperses the new queens one-quarter to one-half mile from their nest, although flights of up to twelve miles have been reported (Tschinkel 1982; Banks et al. 1973). Imported fire ants show no diapause, but seasonal behavior in the United States occurs due to weather.

Division of labor in the colony is largely based on age or size among the workers. Worker size progresses through a continuum, the largest workers weighing fifteen-fold more than the smallest. All workers, except for the largest, begin their adult lives in the interior of the nest, acting as nurses and queen-attendants. As larger workers age, they move to the nest periphery, where they become foragers. Older, smaller workers relay food from foragers to nurses, forage on liquids, and participate in such nest functions as construction, defense, and sanitation.

The mound itself is sturdy and resilent. Plants, such as grasses, frequently grow on the mound, further strengthening it with their roots. Large foraging tunnels extend radially from the mound, branching as they go. Most foraging is done via these underground tunnels and their temporary extensions, so that the presence of the foraging ants is sometimes difficult to detect unless they are quite abundant.

Fire ants are omnivorous, but they prefer to eat other insects and small invertebrates (Tschinkel 1982; Morrill 1978). Like many ants, they also have an affinity for oil-rich seeds (Green 1967) but occasionally patch their nests and foraging tunnels with plant material. Fire ants tend aphids and mealybugs for their honeydew (Green 1952) and are attracted to nectaried cotton (Agnew, Sterling, and Dean 1982).

Damage

Fire ant stings are painful and potentially dangerous. An ant characteristically attacks by pinching the skin with its mandibles while inserting its stinger. Multiple stings by a single ant are common. The venom of *S. invicta* differs from the venom of other stinging ants by its unique composition of

95 percent alkaloids and traces of protein (Blum 1982). Other fire ant venoms, including the venom of *S. richteri*, contain fewer alkaloids. The alkaloids release a histamine, causing local inflammation, itching, and irritation. The protein portion contains enzymes known to be allergens in other hymenopteran venoms (ibid). The sting causes localized burning and itching followed by redness and formation of a pustule six to twenty-four hours later. The pustules, initially sterile, form in response to the action of the alkaloids. If a pustule is broken, such secondary infections as impetigo can set in. If an infection remains undetected, skin grafting, amputation, or incision and drainage may be required (Schmid 1982).

Hypersensitivity rarely occurs, but misdiagnosis of hypersensitive individuals is common. In the elderly, the shortness of breath and shock typical of anaphylaxis may be mistaken for a stroke or heart failure until pustules are found. About 20 percent of the hypersensitive individuals stung for the first time experience anaphylaxis. Generalized allergic reactions from venom enzymes include hives, swelling, nausea, vomiting, and shock. In extreme cases, death may result (ibid).

The emotional and economic impact of fire ant stings is impossible to assess. Reagan Brown, the former Texas Commissioner of Agriculture, estimates that 2.5 million people are stung monthly (Brown 1982). Victims may develop entomophobic reactions, curtailing normal outdoor activities, and medical treatment can be costly: a 1971 physician survey estimated the average cost for treatment at $28.50, with additional home medical expenses (Headley et al. 1982).

One of the most subjective aspects of the fire ants problem in the South has been the pest status of the ant in agriculture. Fire ants are ubiquitous. Virtually every crop grown in the infested area is affected to some degree. However, because the ant is both a general predator and an occasional herbivore, its effects are debatable.

Good cost-benefit analysis for fire ants are nonexistent. Some of the studies that are often quoted are unconvincing. A 1949 report that Alabama farmers were losing $177,000 to $357,000 each year to fire ants was, for example, the result of a poll of farmers in two counties of Alabama (Glancey and Coley 1979). This limited sampling is not reliable, nor is it worthy of extrapolation. Another study has been cited frequently to show that fire ants damage soybeans. In 1976 soybean yields from two fields in Georgia was compared. One field had forty-four ant mounds per acre; the other was mound-free, since it had been treated recently with mirex bait. In that single season, 42.3 bushels per acre were harvested from the infested field, and 44.2 bushels per acre were harvested from the uninfested field. From such a small difference between just two fields in one season, losses to soybean growers in Georgia were estimated at between $170,000 and $1,000,000 per year (Adams et al. 1976). (Other factors now have reduced soybean harvests, for example, combines must lift cutting bars over mounds, leaving more beans in infested fields.) Another study in Florida did a specific cost-benefit study of the control of fire ants in 1975. Sixteen thousand treated acres were compared with controls. The cost of mirex application was $11,947, and the increase in yield in the treated fields was valued at $2,128. The costs outweighed the benefits there by 5.6 to 1.0 (Headley et al. 1982).

Specific damage and benefits to some of the commonly cited crops are summarized in table 13.1. The crops most seriously affected seem to be soybeans and hay (bales left in the field overnight attract hordes of ants, and the mounds may obstruct cutting rigs). Fire ants are most beneficial in cotton, where they prey on boll weevil (Sterling 1978) and *Heliothis* species, and in sugarcane, where they reduce sugarcane borer populations enough to allow farmers to save one to two pesticide applications per season (Maxwell et al. 1982).

One could well agree with the general opinion in the literature before 1957 that fire ants are not very important agricultural pests and may indeed be beneficial overall. The ants are a nuisance in certain settings. But in agriculture, as H. B. Green said in 1952, "the most serious damage done by the imported fire ant is the building of very large mounds. Reports of eighty to ninety mounds per acre in Texas, and up to six hundred mounds per acre "in very severe cases" (Brown 1982), are discredited by Tschinkel (1982), who claims that the upper limit is twenty to twenty-five large, mature colonies per acre. Even allowing for this smaller number, however, the presence of ant mounds can seriously affect cutting and mowing operations with hay, pastures, and soybeans. Since the ant mounds can reach one foot in height and twenty inches in diameter, damage to farm machinery is inevitable if mounds are scattered throughout a field. Crop costs go up when combine blades are broken or when time must be spent repairing equipment.

Efficiency of farmworkers is lost if workers must constantly avoid ants. Time off for medical treatment and medical expenses introduces additional costs. Certain activities are virtually impossible with angry fire ants present, such as removing infested bales of hay left out overnight and making on-the-spot repairs of machinery.

Attacks on newborn calves and other livestock have been recorded but are poorly documented. Fire ants reportedly kill quail, ground squirrels, young deer, and even earthworms (Brown 1982).

Epidemiology, Ecology, and Control

In about 1930, *Solenopsis invicta* appeared in the United States, and like *S. richteri*, it entered through the port of Mobile. At first it was thought to be a red form of *S. saevissima*. Not until 1972 did it achieve separate species status.

Table 13.1 Summary of Damage and Benefits to Crops Resulting from Imported
Fire Ants

Crop	Damage	Benefit
Citrus	Girdling of young trees (Brown 1982)	—
Corn	Damage to germinating plant roots when ants abundant (Glancey and Coley 1979)	—
Cotton	Limited interference with picking; some worker stings (Maxwell et al. 1982)	Predation of boll weevil and *Heliothis* species (Sterling 1978)
Grain Sorghum	Carrying off by ants of poorly covered seed (Maxwell et al. 1982)	—
Hay	Interference of mounds with cutting; infestation of bales left down overnight (Maxwell et al. 1982)	Predation of Alfalfa weevil; deterrence of pea aphid (Morrill 1978)
Nurseries	Damage to pine seedlings (Brown 1982)	Predation of Nantucket Pine-tip moth; reduction of bark beetles (Oliver, Reagan, and Burns 1979; Maxwell et al. 1982)
Okra	Feeding on petals and calyx (little pod damage) (Sterling 1978)	—
Pastures	Interference of mounds with mowing; stings of newborn livestock	Predation of ticks, especially lone star ticks, stable flies, horn flies; reduced tularemia (Summerlin and Kunz 1978; Oliver, Reagan, and Burns 1979)
Soybean	Interference of mounds with harvest; feeding on germinating seeds; some reported but unexplained yield reductions (Adams et al. 1976; Lofgren and Adams 1981)	Predation of three-cornered alfalfa hoppers, stink bugs, several lep. larvae (Maxwell et al. 1982; Sterling 1978)
Sugarcane	Damage to equipment at harvest by mounds (Maxwell et al. 1982)	Predation of surgarcane borer (saves 1–2 pesticide applications per year) (Maxwell et al. 1982)
Sweet Potato	Stings to workers during hand harvest (Maxwell et al. 1982)	Predation of banded cucumber beetle and sweet potato beetle (Maxwell et al. 1982)

Its original range, probably localized around the headwaters of the Paraguay river, is much more restricted than that of *S. richteri*. Once it reached the United States, however, *S. invicta* spread quickly, inhabiting 126 million acres in ten states by 1962.

Because of its arrival twelve years earlier, *S. richteri* may have preconditioned part of the South for *S. invicta* by outcompeting native ant species. Shortly after *S. invicta* appeared in the Mobile area, it displaced *S. richteri* and radiated outward (Tschinkel 1982). By 1940, *S. invicta* had displaced another aggressive introduced species, the Argentine ant (*Iridomyrmex humilus* [Mayr]) (Markin, O'Neal, and Collins 1974). Since then, *Solenopsis invicta* has spread throughout the South, by both natural dispersal and the transport of contaminated nursery stock. The fire ant quarantine of 1956 ended the latter practice, but by the mid-1970s the scattered pockets of fire ants had coalesced (Tschinkel 1982).

The native ant fauna of the United States includes two fire ants, *Solenopsis geminata* (F.) and *Solenopsis xyloni* McCook, the tropical fire ant and southern fire ant, respectively. Both are regarded as inconspicuous members of the southern ant fauna. Compared to imported fire ants they make smaller mounds and are relatively unaggressive (Summerlin and Green 1977).

In the United States, control of fire ants has been attempted almost exclusively with pesticides; the use of biological agents has generally been discounted (Sauer et al. 1982; Jouvenaz, Lofgren, and Banks 1981; Gilstrap 1982). Yet fire ants have many natural enemies in their native South America that may eventually be beneficial in the fight against the pest.

There are several parasitoids attacking the ant in South America, including fourteen species of phorid flies from two genera and many chalcid wasps (Gilstrap 1982). These have never been known to kill an ant colony, however, and so have not been introduced into the United States. Similarly, several organisms attack fire ants in Brazil. These include protozoan, microsporidian, viral, and bacterial diseases (Sauer et al. 1982; Gilstrap 1982). Pathogens exist in the United States but are relatively few and ineffectual, although some microsporidian diseases show promise for possible use against the ant (Gilstrap 1982).

Because there are no reports of biological agents destroying a colony of fire ants, experts believe that they are not efficient enough to have a substantial effect on the pest in this country. Such a conclusion may, however, be erroneous. When dealing with a classic pest species it is proper to talk about the number of individuals killed, but to consider an ant colony as an individual is misleading. Parasitoids kill individual workers. Their effect on a colony is one of attrition. A far better approach than parasitoids would be to apply the principles of biological control of plants to a social insect. To kill a weedy plant, one would import a series of natural enemies: one that feeds on the roots, another that bores into the stem, a leaf-miner, and maybe a budworm. The strategy is to reduce the competitive ability of the plant, so that other plants can displace it. Control of an ant species ought to follow a similar tack (ibid). This approach has been shown to have worked at least once in the case of the Argentine ant in the South (Elton 1958).

The early pessimistic reports about alternative control agents were misleading, as no research station was ever set up in South America to study natural control or to culture parasites and diseases. Only short trips were supported by the USDA (Sauer et al. 1982). When one considers the massive backing of chemical research and chemical control in the United States, the contrast with biological control is dramatic.

Experimentation with cultural control has been almost fruitless. Fire ants survive fire and flooding and most other environmental disturbances with relative ease. Even if its mounds are leveled the colony will continue almost without effect. Fire ants respond to gross disruption of their nest by relocating nearby (Green 1952). In the case of flooding, the ants leave the nest and form a large clump. This ball of ants then floats in the water until it reaches higher ground (Morrill 1974). Flooding has often resulted in very high densities of fire ants at the edges of the flooded area.

One promising development is the use of insect growth regulators (IGRs). Mimics of juvenile hormone, for example, work by disrupting normal metamorphosis of the ants. They have been field tested and have shown fairly good suppression of fire ants (Sauer et al. 1982); however, several applications are needed, and the chemicals are not specific to the fire ant. It seems that widespread use of these IGRs are still some years off.

Although fire ants have probably reduced diversity of ant species in the South by supplanting some native ants, far more harm has been done by the control measures undertaken against them. Two studies show that the various pesticides used in the last twenty years have drastically reduced species diversity. The earlier study concluded that mirex was effective in eliminating oil-feeding species, including fire ants (Markin, O'Neal, and Collins 1974). But the later study found that *S. invicta* increased from 4 percent of all colonies present before mirex treatment to 44 percent one year after treatment (Summerlin, Hung, and Vinson 1977). The results of the studies were hardly a surprise. Ferguson (1972) has said, "With some 200,000 species of organisms in the United States, what are the chances of treating the entire range of a pest insect with a broad-spectrum insecticide and eradicating only the intended target species? In an age of environmental enlightenment, one wonders how a federal agency can entertain such naive and misguided intentions, particularly in view of the abysmal record of success of massive eradication efforts."

History of Eradication Efforts

Pre-Mirex

Beginning in the late 1930s the first report of control of the fire ant were local and small in scale. Highly toxic cyanide dusts and poison baits were applied directly to the mounds and gave immediate relief (Green 1952) but in most cases probably did not kill the colonies.

The first eradication attempt was made by the State Plant Board of Mississippi in 1948. This was in response to concern in several counties that the ant was "devastating the hay industry" (Brown 1961; Mississippi Agricultural Experiment Station 1947–60). The decision to eradicate was probably made in ignorance of the ant's biology and was spurred by an eagerness to use the pesticide chlordane—discovered in 1944 and the first truly potent chemical against ants.

The project ran into several problems and eventually ended without much notice. In 1947–48, the plant board called the program "extensive" and reported that chlordane at a 5 percent dust formulation was "very effective" against fire ants when applied directly to the mounds. Mounds were located visually, and the dust applied from the ground, often after the nest was chopped open with a spade (Green 1952). One year later, the plant board said chlordane was proving "rather satisfactory" against fire ants and initiated a study of the biology of the ant—the first such study reported. By 1949–50, Mississippi was facing the facts more bluntly. The plant board reported, "This new pest is proving very difficult to control." The numbers of ants were reduced by the board's efforts, but few infestations were eradicated. In 1951, with fire ants found throughout the state but in small numbers, the eradication project was terminated (Mississippi Agricultural Experiment Station 1947–60). Total costs to Mississippi were more than $15,000 (USDA 1981).

This first control attempt, though unsuccessful, was instructive. Visual surveys were not good enough, since young colonies were not found. Also, fire ant colonies could survive huge worker mortality without being destroyed. The queen and brood usually managed to escape the pesticides by exiting along the underground foraging tunnels and were unnoticed by the applicators. Thus, the biology of the ant had to be taken into consideration; a potent poison was not enough.

Fire ants were not mentioned again by Mississippi Agricultural Research reports until 1956–57, when the USDA was gearing up its own eradication effort. In fact, in the three states where fire ants were most abundant—Mississippi, Alabama, and Louisiana—the fire ant was seemingly not a very high priority, although both Louisiana and Alabama had been helping farmers purchase chlordane for use against fire ants since 1950 (ibid).

Meanwhile the USDA had become interested in fire ants. A research base was established near Mobile, Alabama, in 1948 to begin a scouting survey to determine the range of the pest and to test various insecticides for use against it. Studies continued until 1953, when research funds were stopped. The USDA then issued a summary of their results in a leaflet (#350), "The Imported Fire Ant, How to Control It." No more fire ant research was funded by the USDA until after the 1957 eradication program was announced (Brown 1961).

The period from 1953 to 1957 is remarkably lacking in news of the fire ant. The 1957 bulletin of the USDA, "Insecticide Recommendations . . . for Control of Insects Attacking Crops and Livestock," for example, did not mention fire ants (Carson 1962). Of the ten southern states then affected, only two listed the ant in the top twenty pest species, and in neither state were fire ants ranked very high (ibid).

Yet in March 1957 the USDA and southern congressmen initiated an appeal to the House Appropriations Committee of Congress for funds to help control fire ants before they "ruin more crops, fields, and lawns, and scare more people" (New York Times 1957a). What caused such grave concern around 1957?

The USDA claims that it responded to an appeal from southern states and Congress to do something about the ant. According to USDA records, however, the only requests were unspecified "inquiries from property owners, organizations, and Congress" about a program against fire ants (USDA 1981). Yet a resolution was adopted to initiate a cooperative federal-state-local program leading to eradication. Considering the relative lack of grave concern about fire ants among the agricultural services of the affected states, a massive eradication campaign hardly seems justified.

One might wonder if the push for eradication was related to the attitudes of USDA policymakers and the political and technological climate of the United States at the time rather than to a great appeal from southern farmers. The mid-to-late fifties was the era of McCarthyism, the incipient Cold War, and massive patriotic spirit. One of the letters read in Congress in support of eradication seems more like a misdirected attack against anything "red" at all: "The government should be building as big a defense against fire ants as they are against the Russians. The ants have already invaded" (New York Times 1957a).

There was also a great confidence among entomologists and farmers in the ability of the new organic insecticides to decimate insect populations. Eradication seemed to be the ultimate American solution to any pest problem. This was the time when the USDA announced its two biggest eradication projects ever—first against the gypsy moth in the Northeast (see chapter 15, this volume), then against the fire ant in the South. The first was large, the second colossal.

In early August 1957 the USDA submitted its final report to Congress. Based on a hasty survey of the range of the ant begun in July 1957, the proposal called for eradication of fire ants from the North American continent. Between 20 and 30 million acres in eight to ten cooperating states would be covered. The total acreage would be sprayed with the insecticides dieldrin and heptachlor at 2 pounds of active ingredient per acre. At the end of August Congress approved $2.4 million per year for the project, to be matched by state funds. This was less than the USDA wanted (USDA 1981).

The first treatments were begun in October, two months before the first major public announcement of the program (*New York Times* 1957a). But only four states appropriated the necessary funds for participation: Alabama, Georgia, Louisiana, and Florida. The USDA responded with a publicity campaign in the South. The *New York Times* reported: "For their ant campaign, members of the Plant Pest Control Division of USDA are utilizing a lesson learned from their battles with the public over the gypsy moth spraying. They have conducted an education program in the South to show the need for spraying. State committees have been organized and the aid of County Farm Agents, sportsmen, and conservationists enlisted. Strong public support has resulted" (Delvin 1957).

By the next year, Texas had joined the program and spraying began. A State-Federal Quarantine Act was passed as well, restricting the movement of materials from areas infested with fire ants. But despite the local support there was plenty of outside opposition.

Within a week of the announcement of the project, Secretary of Agriculture Ezra T. Benson received letters of protest from the National Audubon Society and three prominent Harvard entomologists. The appeals cited the extremely high toxicity of dieldrin to wildlife and birds, as well as the dangers of secondary pest outbreaks from using a broad-spectrum insecticide on such a large scale (*New York Times* 1957a, 1957c). The USDA responded to this early criticism, with calming and vague remarks like the following from M. R. Clarkson, acting Administrator of the Agricultural Research Service: "The project is carefully planned to do the least possible damage to wildlife and other insects" (*New York Times* 1957b), and "A close liaison has been established with the Fish and Wildlife Service of the Department of the Interior and the states involved" (Brown 1961).

In fact, the project was begun in haste and without any provision for monitoring the effects on human health and wildlife. Only after the first year of spraying were any studies initiated to examine the side-effects (ibid); the first meeting between the USDA and the Department of Fish and Wildlife occurred one month after spraying had already begun (ibid). The main reasons for this were the insistence of Congress that the program begin at once and the relative lack of research funds provided (Shapley 1971). Congress was presumably acting on advice and reports from the USDA when making this decision; blame cannot be placed on one group.

When the environmental impact reports did come in, they were devastating. The chemicals chosen to do the job were the major problem. Robert Metcalf et al. (1982) stated succinctly: "The insecticides chosen for control and later eradication efforts . . . can be collectively described as extremely persistent in the environment, broad-spectrum in their toxic action, highly bioaccumulative with long residence times in human and animal tissues, and generally hazardous to a wide variety of terrestrial and aquatic wildlife. These insecticides have also been shown . . . to be chemical carcinogens."

A study in Louisiana on four farms three weeks after treatment with heptachlor at 2 pounds per acre found the following animals dead: 53 mammals, 222 birds, 22 reptiles, many fish and frogs—a combination of more than 60 species. Ninety-five percent of the corpses were autopsied, and all contained heptachlor (Hinkle 1982).

By the end of 1958, support for this spraying effort began to wane. Alabama Senator Sparkman and Representative Boykin cosponsored a bill in the House and Senate to stop the spraying until the benefits and dangers could be assessed (Brown 1961). The measure failed to pass. Meanwhile the USDA opened a new laboratory in Gulfport, Mississippi, specifically for improvement of methods. In 1959, the rate of application was lowered to 1.25 pounds of heptachlor per acre.

The turning point was 1960. The Alabama legislature refused to appropriate matching funds for the project. Florida held hearings on the pest status of fire ants and decided to withhold matching funds. The Florida plant commissioner announced that "efforts to stamp out fire ants permanently in Florida ha[d] failed" (ibid). In a last gasp, the USDA offered heptachlor free to landowners in Texas in exchange for absolving the state and federal governments from any damage that might result from its use. But Texas, too, withheld matching funds for eradication of the fire ant. Finally, the Food and Drug Administration reduced tolerances for heptachlor on pastures to zero when it was found that heptachlor degrades to an even more toxic and persistent epoxide appearing in adipose tissue of mammals (Metcalf 1982).

In 1961, the third year of spraying, the dose was lowered to 0.25 pounds of heptachlor per acre, applied twice. It was also found that fire ants were beginning to reinfest the areas sprayed one to two years before, and at much higher densities (Brown 1961). Spraying with heptachlor was completely stopped in 1962.

This initial phase of the eradication attempt by the USDA lasted four years, cost about $15 million, and covered 2.5 million acres with a highly toxic, persistent insecticide. Only temporary relief was obtained, and meanwhile the fire ant increased its range from 90 million acres in 1957 to 126 million acres in 1962. Wildlife in the spray zone was greatly affected; for example, the number of birds sighted in sprayed areas in Texas and Louisiana decreased by 85 percent (Hinkle 1982).

Eradication was not abandoned, though. In 1960, when the fate of heptachlor was becoming obvious to everyone, a related compound called chlordecone (Kepone) was found to be effective against fire ants, and mirex, a near twin of Kepone, was found to provide control as good as heptachlor while being less toxic to wildlife. The Gulfport laboratory formulated mirex into a corn-grit and soybean oil bait, which acted as a delayed stomach poison and had some target selectivity.

Mirex bait was unveiled to replace heptachlor in 1962. It was hailed as the ultimate solution to the problems of the first attempt. Secretary of Agriculture Orville Freeman proclaimed, "Mirex has no harmful effect on people, domestic animals, fish, wildlife, or even bees; and it leaves no residue in milk, meat, or crops" (Shapley 1971). So, almost without missing a beat, the next phase of the eradication attempt began, still without any clear examination of possible alternatives.

Mirex

Mirex possesses a number of qualities that made it an ideal control agent in the pre–*Silent Spring* era but that made it controversial soon after it was introduced. Mirex is a highly stable hydrocarbon with a tight cage structure. Studies have demonstrated that it bioaccumulates in adipose tissue and is biomagnified through food chains.

Mirex was first patented in 1954 by Allied Chemical and Dye Corporation for use as an insecticide (Kaiser 1978). In the 1960s various chemical companies obtained patents specifying other uses. At first mirex was marketed as a rodenticide and a smoke-generating compound. Hooker Chemicals and Plastics—later to become infamous as contaminators of Love Canal in New York—patented mirex as a flame retardant, which continues to be its most common use. Mirex bait was first used against fire ants in 1962, when it replaced heptachlor.

The introduction of mirex bait brought great optimism into the USDA project. Criticism from conservationists lessened, and there was renewed confidence that fire ants could be eradicated.

The optimism was short-lived. Despite large-scale applications of mirex, initially at 10 pounds per acre but reduced to 2.5 pounds per acre by spring 1963, immature fire ant colonies were not eliminated. Skepticism about the eradication effort again began to spread, and the political debate rekindled. In the House of Representatives, Jamie L. Whitten, chairman of the Subcommittee on Agriculture and an avid believer in eradication, tried to avert impending budget cuts to the fire ant program by threatening a retaliatory assault on the entire USDA operating budget (Shapley 1971). Nevertheless, the General Accounting Office, a government watchdog agency wary of the program and critical of the USDA for squandering government funds and ignoring scientific opinion (Hinkle 1982), blocked funding, so that in 1965 and 1966 no budget was produced for fire ant eradication.

During this hiatus, Whitten and Senator Spessard L. Holland (D-Florida), head of the Subcommittee on Agriculture of the Senate Appropriations Committee, pushed for federal subsidization for a massive new eradication program. Both Senate and House committees included a substantial number of southern members. They were accompanied by southern agricultural commissioners who lobbied by showing photographs of sting victims encrusted with pustules. Even Richard M. Nixon got involved, promising an effective fire ant control as he campaigned in the South for the 1968 election. (Whitten's propesticide stance was later called into question when the *Washington Post* revealed connections between three pesticide manufacturers and Whitten. The connection concerned Whitten's book *That We May Live,* published in 1966 [Shapley 1971]).

In late 1966, the Southern Plant Board convinced the USDA and Congress to approve a three-year eradication program. The program included a survey to assess the effectiveness of mirex bait, two to four annual treatments at 1.25 pounds per acre, which amounts to less than a teaspoon per acre of active ingredient (USDA 1981). An electronic guidance system was perfected for massive aerial applications.

In 1967, a Senate subcommittee requested that the USDA determine the feasibility of fire ant eradication using mirex bait. The USDA and the states of Georgia, South Carolina, Florida, and Mississippi initiated cooperative tests, some covering more than two million acres. Mirex bait suppression rates were nearly 100 percent against fire ant mounds and foraging workers in studies by Markin, O'Neal, and Collins (1974). In another study, incipient colonies were still found after three applications of mirex bait, either from queens surviving within sites or entering from the periphery. The principle investigators offered vague recommendations: "The results did not conclusively prove that [mirex bait] can be used to eradicate the red and black fire ants," and later, "Total elimination from large isolated areas may be technically feasible" (Banks et al. 1973).

Despite the equivocal results, an updated and expanded eradication program resulted, targeting the entire 126 million acres known to be infested. The USDA and participating states were to split the $200 million cost over twelve years (Metcalf 1982). Meanwhile, an independent report issued from the National Academy of Science dismissed the program as biologically and technologically impossible (Mills 1967).

The eradication program began in 1967 with large-scale treatments of areas in Florida, Georgia, and Mississippi. Coordination of efforts was poor among the nine states involved in the program. State funding was not always available, and some states defaulted on their payments. A USDA official commented, "One year the states would come to us filthy rich with money for fire ants and the next minute they wouldn't have a dime" (Shapley 1971). Eradication seemed a futile prospect, since treated states were inevitably invaded by fire ants from adjacent, untreated states.

As the program gained momentum, it also picked up a number of critics whose primary concern was mirex. The Fish and Wildlife Division of the Department of the Interior had been a foe of heptachlor and was no happier with mirex. The division and the newly formed Environmental Protection Agency presented a substantial obstacle to the program in 1970.

Suspicions about the dangers of mirex arose in the 1960s from tests involving fish, birds, and rodents. In 1969, mirex was the fourth most abundant pesticide residue found in mollusks residing in coastal estuaries around the Atlantic, the Gulf of Mexico, and Pacific (Hinkle 1982). Evidence of persistence and biomagnification mounted in the years of heavy mirex usage. Residues showed up in aquatic plants, crayfish, birds, fish, turtles, "animals normally raised for human consumption," and humans (Waters, Huff, and Gerstner 1977). When labeled mirex was administered to rats (6 mg/kg), 34 percent was stored in body tissues, mostly in fat cells, muscle, and liver. In pregnant rats, mirex entered the placenta and accumulated in the fetuses (ibid). In addition, excessive doses of mirex (10 mg/kg over 21 days) induced hepatomas in 45 percent of experimental mice. Control mice sustained a 4 percent hepatoma incidence. Significant residues were detected in human adipose tissue in a study begun in 1971 (Kutz et al. 1974).

In 1970, the Department of the Interior, encouraged by Secretary Walter Hickel, restricted mirex and curtailed its use on Interior lands (Shapley 1971). Then in August 1970, the Environmental Defense Fund (EDF) filed a motion to stop mirex spraying and to halt the program altogether. Spokesmen for Allied Chemical and the USDA avoided commenting on the suit (*New York Times* 1970). The media in the northern states glibly covered the controversy, poking fun at the intense dislike shown by southerners toward such an insignificant creature. The *New York Times* (Nordheimer 1970) blamed the ant for causing "discomfort to picnickers and agricultural workers."

As a result of the EDF suit, a detailed environmental impact statement (EIS) was requested. Within two months, the USDA submitted the EIS, admitting grave logistical and financial problems. At this juncture, the USDA abandoned their grandiose eradication scheme, adopting a more realistic "control" program. Bait applications on such sensitive areas as estuaries, game refuges, and bodies of water would be avoided, but problem areas would be aerially treated (USDA 1971).

In March 1971, less than two months after the EIS appeared, the EPA publicized its intention to cancel the use of mirex, although the actual cancellation order was postponed for the next three years while the EPA and Allied Chemical engaged in courtroom battles. Overwhelming evidence pointed to the dangers of mirex to wildlife and humans. The degradative products of mirex, photomirex and Kepone, are more dangerous than mirex itself. Photomirex residues have been detected in herring gull eggs and fish (Kaiser 1978). Kepone, structurally similar to mirex, was linked to reproductive problems and nervous system disorders in exposed factory workers. The National Cancer Institute reported that a high proportion of Kepone-fed mice suffered from liver cell carcinomas (Metcalf 1982). Such discoveries, as well as the many examples of mirex bioaccumulation and widespread contamination by polychlorinated biphenyls, finally led to the banning of mirex.

In early 1976, Allied announced that it would stop manufacturing mirex and sold its factory and rights to the state of Mississippi for one dollar (Marshall 1982). Mississippi continued to spray mirex aerially until the end of 1977. By mid-1978, mirex was no longer permitted for ground use or mound applications (Hinkle 1982). The fire ant control program lost not only mirex, but much of its following as well. The Office of Management and Budget reduced the annual appropriations ten-fold, from $9 million to $900,000. But Congress rescued some of the funds for development of methods, monitoring, quarantine, and control (ibid).

Meanwhile, in Mississippi, chemists at the old mirex factory were brewing a new ant poison. Subsidized by earnings from last-minute sales of mirex, the chemists developed ferriamicide, which they touted as being biodegradable. The EPA immediately granted a six-month emergency use permit. Before spraying began, ferriamicide met with trouble: the EDF filed a suit against the EPA claiming tests on degradation and mutagenicity were incomplete. For a short time, ferriamicide was used only for mound treatment, while Mississippi argued that the compound was perfectly safe.

Like its predecessor, mirex (ferriamicide is 0.05 percent mirex), ferriamicide yielded photomirex and Kepone by-products. Addition of amine and ferrous chloride decreased the half-life from 12 years to 0.15 years, and less Kepone was produced by adding propylene glycol as a stabilizer (Alley 1982). Also, ferriamicide was 10 to 20 times cheaper to manufacture and apply than other chemicals. Yet by mid-1982, the prospects for ferriamicide looked dim. The Hazard Evaluation division of the EPA said the half-life estimations of ferriamicide had been exaggerated. The EPA memo warned that "as much as 40 percent, or 94 percent including degradates, may remain after three years" (Marshall 1982). Today the future for ferriamicide seems uncertain.

Post-Mirex

Even after three disasterous attempts, the USDA continued to plot against fire ants on a grand scale. In 1978, it began testing a new chemical called hydramethylnon (Amdro), formulated in a bait similar to the mirex bait. Amdro acts as a delayed stomach poison but is not a chlordane derivative like the previously used chemicals. Results were highly favorable. Amdro gave very good control of fire ants and yet had a mammalian toxicity comparable to malathion.

Amdro did have some problems. It did not achieve kill in the 90 percent range as did mirex, but since eradication was no longer the stated goal, this was acceptable. Amdro was toxic to fish, though almost insoluble. Since the project did not plan to spray over bodies of water, this too was an acceptable drawback.

So, in 1979, after extensive field testing of Amdro bait against fire ants, the EPA registered it for use on pastures but not on crops. The USDA quickly developed a new proposed federal-state-local cooperative program to control the ant based on aerial and ground applications of Amdro. In its environmental impact statement, the USDA praised Amdro: "The toxicant degrades rapidly in the environment in sunlight with a half-life of less than 24 hours. It is only slightly soluble in water . . . does not leach in the soil and is degraded by microorganisms. . . . the use of Amdro for the control of fire ants will not result in biological magnification of Amdro-related residues in the environment" (USDA 1981).

The new proposal was essentially the same program as had been in effect before the demise of mirex, with the same addendum that eradication was not feasible at this time (USDA 1981).

By the end of 1980, one-half million acres had been treated; the plan called for treatment of one million acres each year. Support for the renewed fire ant control program has already sprung up. In May 1982 the Southern Legislative Conference adopted a resolution to urge southern legislatures to appropriate the necessary state funding (Brown 1982).

Approval of a new compound by the EPA in March 1983 may challenge the popularity of Amdro and other pesticides currently in favor. Dimethylnonane (Pro-Drone), manufactured by Stauffer Chemical, is an insect growth regulator that interferes with caste development. Development of worker ants is suppressed, and the colony, its internal organization thus disrupted, eventually succumbs to starvation. Stauffer estimates that 80 percent to 90 percent mortality. Critics complain that Pro-Drone is much more expensive than ferriamicide or Amdro and is much more difficult to apply (Marshall 1982).

The future of fire ant control is unclear. Mississippi continues to push ferriamicide. American Cyanamid is intensifing its production of Amdro. Insect growth regulators, relatively selective and effective, face an uncertain future.

Evaluation

Several points relate specifically to consideration of an eradication effort. First, fire ants may have already reached the limits of their range in the United States. Cold winter temperatures have stopped their northward spread, and the need for warm rain may prevent their westward movement. Texas west of the hundredth meridian may be too dry, and California, which lacks summer rains, may be too dry in the warmer months.

Yet, heavily irrigated agricultural valleys in California could provide an ideal habitat for fire ant establishment (S. J. Risch, personal communication, 1985). It is conceivable that fire ants will reach the west coast by themselves or through human activity, or they may in time adapt to drier conditions.

Second, mature colonies are territorial. Within an area around its nest, a colony will not tolerate ants from another colony and will kill newly mated queens attempting to found a nest. As a result, fire ant populations have a self-imposed carrying capacity of approximately twenty-five mature mounds per acre (Tschinkel 1982). Counts higher than this indicate young colonies in the growth phase.

Third, ants depend on pheromones for the maintenance and coordination of colony growth and activity. Five pheromones have been identified and may provide an avenue for specific control. Everything from locating food to maintaining caste structure could be interfered with it these chemicals could be synthesized and used to confuse the ants. Insect growth regulators are also promising control prospects. Juvenile hormone, for instance, is vital to the proper growth of the colony: when the balance of juvenile hormone in a nest is disrupted, the colony can be critically affected.

Fourth, fire ants are weedlike, adapted for fast colonization of disturbed habitats and fast dispersal. Cooperative nest excavation makes the early stage of the colony much less vulnerable; cooperative nests are on the average three time larger than those founded by a single queen during the initial growth phase (ibid). Since most natural mortality occurs during the first month of colony growth, this gives fire ants a great advantage over competing species that do not cooperate. Fire ants have a very high fecundity. Each colony produces approximately 4,500 new queens each year—

about 100,000 new queens per year per acre (Maxwell et al. 1982). This is between 30 percent and 40 percent of the total production of the colony (Tschinkel 1982). But the weedy nature of the fire ant is also limiting; they are poorly adapted to climax communities and have much less success invading areas where other ant species are established. Fire ants depend on disruption for their success.

To be successful, eradication efforts must be aimed specifically at fire ants and must avoid being generally disruptive to the ecosystem. In principle, a nonspecific myrmecocide, though it would kill fire ants along with other ants, would also facilitate the reinfestation of fire ants by eliminating its competitors. This has been demonstrated (Buren et al. 1974). Using a model, it is predicted that an ant fauna consisting of 1 percent fire ants will shift to 99 percent fire ants in four years given a 95 percent annual mortality rate of all ant species. Also, the extremely high fecundity and rapid maturity of the ant make it vital that control measures be highly lethal. One USDA study estimated that a kill of 99.9 percent to 99.99 percent per spray would be necessary in a repeated spray program in order to exterminate fire ants (Lofgren and Weidhaas 1972).

It has been twenty-six years since the USDA began mass spraying to eradicate fire ants. About $200 million has been spent. Yet, the fire ant problem is worse now than it ever has been. Many of the native southern ants that once competed with fire ants have been greatly reduced. The treated areas represent a highly disturbed habitat, ideal for fire ant reinfestation.

Ironically, it was only after the situation had been so badly mishandled that the question of the feasibility of eradication was broached and answered in the negative (van den Bosch 1978). If the biology of the ant had been considered in the first place, it is unlikely that the environmental poisons, heptachlor and dieldrin, would have been considered. The very idea of using a broad-spectrum pesticide on a species so well adapted to colonizing disturbed areas would have seemed laughable. If the USDA's education and eradication project had not already been fired up, it is unlikely that politicians would have been so eager to keep the wheels turning with mirex when the initial program stalled.

But the most disturbing question that remains unanswered is, Why did the whole thing begin in the first place? Fire ants were a relatively benign target for so massive a mobilization of a government agency. Why have taxpayers spent over $200 million trying to rid themselves of a little southern discomfort?

The answer may be at least partly unrelated to entomology, agriculture, or economics. At some point, in the eyes of politicians and USDA administrators, fire ants attained an undeserved and unrealistic image. They became the Great American Pest and, considering the political climate of the time, might better have been called The Great Red Menace. Given this perspective, one might say that America, represented by the USDA, was trying to teach all pests a lesson, in effect: "Stay out of my backyard."

In one sense we can be thankful that so tenacious a species was the target. By steadfastly refusing to succumb, fire ants forced a reexamination of eradication philosophy. But the costs of that lesson have been considerable, both in dollars and in harm done to the environment.

Literature Cited

Adams, C. T., J. K. Plumley, C. S. Lofgren, and C. S. Banks. 1976. Economic importance of the red imported fire ant, *Solenopsis invicta* Buren (1): Preliminary investigations of impact on soybean harvest. *J. Georgia Entomol. Soc.* 11:165–69.

Agnew, C. W., W. L. Sterling, and D. A. Dean. 1982. Influence of cotton nectar on red imported fire ants and other predators. *Environ. Entomol.* 11:629–34.

Alley, E. L. 1982. Ferriamicide: A toxicological summary. In *Proceedings of the symposium on the imported fire ant,* June 7–10, Atlanta, Georgia, ed. S. L. Battenfield, 144–53.

Banks, W. A., B. M. Glancey, C. E. Stringer, D. P. Jouvenaz, C. S. Lofgren, and D. E. Weidhaas. 1973. Imported fire ants: Eradication trials with Mirex bait. *J. Econ. Entomol.* 66:785–89.

Blum, M. S. 1982. Chemistry and properties of fire ant venoms. In *Proceedings of the symposium on the imported fire ant,* June 7–10, Atlanta, Georgia, ed. S. L. Battenfield, 116–18.

Brown, R. V. 1982. The State perspective of the imported fire ant. In *Proceedings of the symposium on the imported fire ant,* June 7–10, Atlanta, Georgia, ed. S. L. Battenfield, 36–39.

Brown, W. L. 1961. Mass insect control programs: Four case histories. *Psyche J. Entomol.* 68:75–111.

Buren, W. F., G. E. Allen, W. H. Whitcomb, F. E. Lennartz, and R. N. Williams. 1974. Zoogeography of the imported fire ants. *J. N.Y. Entomol. Soc.* 82:113–24.

Carson, R. 1962. *Silent spring.* New York: Fawcett Crest.

Delvin, J. C. 1957. U.S. helping South in fire ant fight. *New York Times,* December 23, 25.

Elton, C. S. 1958. *The ecology of invasions by animals and plants.* New York: John Wiley and Sons.

Ferguson, D. E. 1972. Fire ants and the eradication concept. In *Environmental Quality 2,* ed. H. L. Sherman, 4: 21, 22–71. Columbus, Miss.: Mississippi College for Women.

Gilstrap, F. E. 1982. Biological control: Its historic use vs. prospective value to control the imported fire ant. In *Proceedings of the symposium on the imported fire ant,* June 7–10 Atlanta, Georgia, ed. S. L. Battenfield, 214–25.

Glancey, B. M., and J. D. Coley. 1979. Damage to corn by the red imported fire ants. *J. Georgia Entomol. Soc.* 14:198–201.

Green, H. B. 1952. Biology and control of the imported fire ant in Mississippi. *J. Econ. Entomol.* 45:593–97.

Green, H. B. 1967. The imported fire ant in the southeastern United States. *Bull. Mississippi Agric. Exp. Sta.* 737.

Headley, J. C., A. Aspelin, C. T. Adams, T. Brooks, R. E. Brown, G. A. Carlson, C. Carr, F. James, Q. Jenkins, and J. Schaub. 1982. Socio-economic factors relating to the fire ants and its management. In *Proceedings of the symposium on the imported fire ant,* June 7–10, Atlanta, Georgia, ed. S. L. Battenfield, 41–50.

Hinkle, M. K. 1982. Impact of the imported fire ant control programs on wildlife and quality of the environment. In *Proceedings of the symposium on the imported fire ant,* June 7–10, Atlanta, Georgia, ed. S. L. Battenfield, 130–43.

Jouvenaz, D. P., C. S. Lofgren, and W. A. Banks. 1981. Biological control of imported fire ants: a review of current knowledge. *Bull. Entomol. Soc. Amer.* 27:203–8.

Kaiser, K. L. E. 1978. The rise and fall of Mirex. *Environ. Sci. and Technol.*: 520–28.

Kutz, F. W., A. R. Yobs, W. G. Johnson, and G. B. Wiersma. 1974. Mirex residues in human adipose tissues. *Environ. Entomol.* 3:882–84.

Lofgren, C. S. 1986. History of imported fire ants in the United States. In *Fire ants and leaf-cutting ants: Biology and management,* ed. Lofgren C. S. and R. K. Vander Meer, 36–47. Boulder and London: Westview.

Lofgren, C. S., and C. T. Adams. 1981. Reduced yield of soybeans in fields infested with the red imported fire ant, *Solenopsis invicta* Buren. *Fla. Entomol.* 64:199–202.

Lofgren, C. S., and D. E. Weidhass. 1972. On the eradication of imported fire ant: a theoretical appraisal. *Bull. Entomol. Soc. Amer.* 18:17–20.

Markin, G. P., J. O'Neal, and H. L. Collins. 1974. Effects of mirex on the general ant fauna of a treated area in Louisiana. *Environ. Entomol.* 3:895–98.

Marshall, E. 1982. Mississippi Inc., pesticide manufacturer. *Science* 218:548–50.

Maxwell, F. G., W. A. Banks, J. L. Bagent, W. L. Buren, O. Franke, S. J. Risch, A. Sorenson, W. L. Sterling, and J. L. Stimac. 1982. Population dynamics of the imported fire ant. In *Proceedings of the symposium on the imported fire ant,* June 7–10, Atlanta, Georgia, ed. S. L. Battenfield, 67–74.

Metcalf, R. L. 1982. A brief history of chemical control of the imported fire ant. In *Proceedings of the symposium on the imported fire ant,* June 7–10, Atlanta, Georgia, ed. S. L. Battenfield, 125–29.

Metcalf, R. L., D. L. Severn, E. L. Alley, M. Blum, M. K. Hinkel, H. N. Nigg, W. H. Schmid, and J. Wood. 1982. Environmental toxicology. In *Proceedings of the symposium on the imported fire ant,* June 7–10, Atlanta, Georgia, ed. S. L. Battenfield, 75–81.

Mills, H. B. 1967. *Report of committee on imported fire ant.* National Academy of Science to Administrator, Agricultural Research, USDA. Washington, D. C.: Government Printing Office.

Mississippi Agricultural Experiment Station. 1947–60. *Annual Reports.* Mississippi Agricultural Experiment Station.

Morrill, W. L. 1974. Dispersal of red imported fire ants by water. *Fla. Entomol.* 57:39–42.

———. 1978. Red imported fire ant predation on the alfalfa weevil and pea aphid. *J. Econ. Entomol.* 71:867–68.

New York Times. 1957a. Aubudon Society calls war on fire ant chemical peril to humans and wildlife. *New York Times,* December 25, 33.

―――. 1957b. Three experts protest fire ant program. *New York Times,* December 27, 11.

―――. 1970. Suit seeks to curb fire ant pesticide in nine southern states. *New York Times.* August 6, 30.

Nordheimer, J. 1970. Environmentalists fight U.S. spraying plan on fire ants in the South. *New York Times,* December 13, 78.

Oliver, A. D., T. E. Reagan, and E. C. Burns. 1979. The fire ant: An important predator of some agricultural pests. *Louisiana Agriculture* 22:6–7, 9.

Sauer, R. J., H. L. Collins, G. Allen, D. Campt, T. D. Canerday, G. Larocca, C. Lofgren, G. Reagan, D. L. Shankland, M. Trostle, W. R. Tschinkel, and S. B. Vinson. 1982. IFA management strategies. In *Proceedings of the symposium on the imported fire ant,* June 7–10, Atlanta, Georgia, ed. S. L. Battenfield, 91–110.

Schmid, W. H. 1982. The medical aspects of the imported fire ant. In *Proceedings of the symposium on the imported fire ant,* June 7–10, Atlanta, Georgia, ed. S. L. Battenfield, 119–21.

Shapley, D. 1971. Mirex and the fire ant: Decline in fortunes of "perfect" pesticide. *Science* 172:358–60.

Sterling, W. L. 1978. Imported fire ant . . . may wear a gray hat. *Texas Agric. Progress* 24:19–20.

Summerlin, J. W., and L. R. Green. 1977. Red imported fire ant: A review on invasion, distribution, and control in Texas. *Southwestern Entomologist* 2:94–101.

Summerlin, J. W., A. C. F. Hung, and S. Bradleigh Vinson. 1977. Residues in nontarget ants, species simplification and recovery of populations following aerial applications of mirex. *Environ. Entomol.* 6:193–97.

Summerlin, J. W., and S. E. Kunz. 1978. Predation of the red imported fire ant on stable flies in pastures in east Texas. *Southwestern Entomologist* 3:260–62.

Tschinkel, W. R. 1982. History and biology of fire ants. In *Proceedings of the symposium on the imported fire ant,* June 7–10, Atlanta, Georgia, ed. S. L. Battenfield, 16–35.

U. S. Department of Agriculture (USDA). 1971. Imported fire ant cooperative federal-state control and regulatory treatment. CEQ 10759. Washington, D.C.: USDA Forest Service.

―――. 1981. Cooperative imported fire ant program. *Final programmatic environmental impact statement.* USDA-APHIS-ADM-81-01-F. Washington, D.C.: USDA.

van den Bosch, R. 1978. The pesticide conspiracy. New York: Doubleday.

Waters, E. M., J. E. Huff, and H. B. Gerstner. 1977. Mirex: An overview. *Environ. Res.* 14:212–22.

14

Date Palm Scale

Kent M. Daane and Larry R. Wilhoit

Date palms, *Phoenix dactylifera* L., were cultivated in Mesopotamia as long ago as 3000 B.C. and possibly earlier, making them one of the first plants domesticated by humans. In the subtropical region of the Middle East and North Africa, dates have been a staple food source for thousands of years. Americans know of date palms primarily for their fruit, but the tree provides much more: the sap can be made into an alcoholic drink, the trunk is excellent wood for construction, the fiber can be used to make ropes and mats, the leaflets for baskets, and the seeds for charcoal. Dates require a high temperature for fruit production and need low rainfall and humidity, with plenty of ground water for their roots (Cockerell 1907; Popenoe 1973).

The Spaniards brought date palms to the New World in the late 1700s and started plants in their California missions, but the coastal areas where they settled were unsuitable for date production. Plants later were started in the interior desert regions of California and Arizona, where conditions for date palms were much better. All the early palms, however, were grown from seed, and because they do not breed true, the fruit was variable and usually unsatisfactory. Also, since dates are dioecious, half of the plants produced no fruit at all. Thus in 1890 the United States Department of Agriculture (USDA), sensing the potential for a new, productive food crop, imported palm offshoots from Algeria and Egypt to obtain standard, good-bearing varieties. Offshoots are vegetative growths from a bud usually at the base of the trunk of the parent tree. These offshoots can be planted, producing a tree nearly identical to the parent tree. Imported offshoots were started in experimental plots in Arizona and California, and in the next few decades hundreds of offshoots were imported. As standard plantations became established, many of the seedling plants were abandoned and became weeds. Later, these abandoned palms were to play a role in the eradication of *Parlatoria* date scale. It was not until 1920 that the commercial development of the date industry was solidly established. Climate limited the industry to small areas in Arizona, California, and Texas, with the Coachella Valley of California becoming the primary date-growing region and producing most of the dates in the United States. These small areas were excellently suited for the date palm, and in a short time date production in the United States was twice that of the Middle East. The fruit held the promise of being a very profitable commercial crop (*New York Times* 1928; Nixon 1959; Nixon 1971).

This promise was threatened by a quiet menace, the date palm scale, *Parlatoria blanchardi.* The insect coevolved with the date palm in its native region, where it is not considered a major pest (fig. 14.1). When it came to the United States on those first offshoots, arriving without its associated natural enemies, it was soon elevated from minor to major pest. As more date palms were planted the scale increased in number and economic importance. It was soon evident that a commercial industry in the United States was impossible unless control measures were taken.

Life History and Epidemiology

Parlatoria blanchardi (Targioni-Tozzetti) (Homoptera: Diaspididae) is an armored scale, elongate and oval in shape. The female is about one-twen-

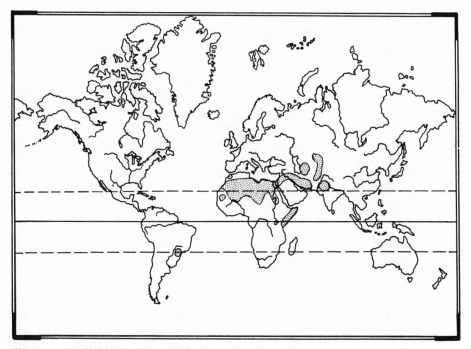

Figure 14.1 Worldwide distribution of the date palm scale, *Parlatoria blanchardi* (adapted from Commonwealth Institute of Entomology 1962).

tieth of an inch long and one-thirty-fifty of an inch wide, whereas the male is slightly smaller. On the palm the female appears as a dark speck, and the male is completely white because of the accumulation of shed skins as the insect molts (Boyden 1941; Cockerell 1907; Smirnoff 1957). Eggs are laid under the covering of the female. The newly hatched nymphs have well-developed eyes, legs, and antennae. Sex can be determined in the early instars by the presence of a prominent spine on the outer surface of the tibia of all six legs of the male, which is lacking on most females (Stickney 1934). To reach the adult stage the female molts twice and is wingless and sessile as an adult. The male is usually winged. Stickney (1934) has provided a detailed description of the insect's anatomy, separating instars by the number and position of pore plates, setae, and lobes and describing mouthparts, head, and thorax.

The biology is similar to other diaspin scales. There are features of this group, and this species in particular, however, that become important in the eradication effort. In the United States the scale has relatively few hosts, and the date palm is the only commercially important tree. The scale also survives on the Canary Island palm, which was treated with equal importance in the eradication program. Two other palms, California fan and Doum, could support small populations but were not suitable hosts and required constant reinfestation (Boyden 1941).

The date scale will spread over the foliage and fruit, damaging both the tree and the value of the crop. In heavy infestations the leaves die back, but the palm rarely dies. The structure of the tree provides excellent protection for the scale. Each leaf that emerges has a triple-layered sheath that encircles the growth center. At first, the sheath is white and succulent but later becomes tough, brown, and fibrous. The leaves, which will be ten to twenty feet when full grown, are attached to the trunk in whorls by tough fiber bands. The leaf bases along with the protective sheath form a thick trunk covering that persists after the leaf blades are removed. Small populations of scale will develop here, protected by the fibrous leaves. There is probably no better protected habitat for any scale than under the leaf bases of date palms (Boyden 1941; Smirnoff 1957).

As the female is wingless and motile only briefly in the preadult crawler stage, dissemination is slow. The crawlers usually settle close to the oviposition site, which intensifies the economic damage but also retards dissemination. Invasion of a new area is accomplished by direct contact between palms, by the wind, by rodents and birds, and by human activities during pollination and harvest (Boyden 1941). Spread from continent to continent and from state to state is by infested offshoots. Transport of the scale within growing regions is very slow, usually less than one and a half miles per year, most of which has been attributed to the wind (Cockerell 1907). Because of the limited host range and growing region of the scale, quarantine is a feasible control technique.

In addition to the slow spread of the scale, another aspect of the pest's biology making control and even eradication promising is its surprisingly low fecundity. The largest number of eggs and nymphs reported under a single female was between seven and eleven, depending on region. Because the scale is multivoltine, however, populations can build to large numbers rather quickly. There are usually three to five generations a season before a short winter diapause (Bartlett et al. 1978; Smirnoff 1957). The increase in scale numbers corresponds with the development of the fruit, increasing the economic damage.

Control

In the Middle East and areas close to the origin of the scale, constant reinfestation precludes eradication. In these regions biological and chemical control have been used with varying degrees of success.

Examples of insecticides currently used are diazinon, methyl parathion, carbaryl, dimethoate, and oils (Butan 1975; Siddig 1975). Although these insecticides reduce scale populations, the procedure is not unlike the defoliation and torch method used in the United States in the 1920s and 1930s, except that the insecticides have replaced the flame. To reach the protected scale the leaf bases are removed, which in itself leads to a significant reduction in the scale population. The advantages the insecticides have over the torch are that the crop is not lost immediately after treatment and application is much easier.

Although insecticides have been effective, they also present a repeated cost; where biological control has been successful it is without such costs. *Parlatoria blanchardi* was described and recognized as a pest in the early 1890s, when date palms were imported into the United States. Since that time natural enemies have been imported and established in numerous North African and Middle Eastern date palm orchards. In Algeria a severe outbreak caused the death of 100,000 date palms in the early 1900s. Two predators, *Cybocephalus palmarum* Peyerimhoff and *Pharoscymnus numidicus* (Pic.), were introduced in 1925, and although unable to completely control the scale, these predators brought pest populations down to manageable levels. From Algeria the date palm scale spread into Morocco, and from 1950 to 1954 more than dozen predator species were collected and released into infested areas of Morocco. Wherever these predators became established, pest populations were lowered or destroyed. In 1957 a thorough study of the natural enemies showed that a complex of thirty-five predators controlled the scale in North Africa (Smirnoff 1957). This study also brought out the importance of proper timing and selection of natural enemies in each desert microhabitat. From these and other studies it appears that biological control is an effective alternative to insecticidal treatments. But in many instances, growers do not consider this alternative because of a bias toward chemical control and a desire to have orchards clean of any insects.

History of Eradication Efforts

The eradication attempt in the United States can best be divided into two periods: an initial attempt from 1914 to 1927 that failed and a second one from 1927 to 1936 that succeeded.

As mentioned, the scale was first introduced into the United States on offshoots from Algeria and Egypt in 1890. It can be assumed that some of the other shipments that arrived over the next twenty-two years were also infested. Once introduced, populations were left unchecked for fifteen years while the experimental and commercial orchards became established. During this time the scale spread into the Imperial and Coachella Valleys of California, into the Yuma and Salt River Valley districts of Arizona, and near Laredo, Texas (Boyden 1941). It became quite evident that unless some control measure was initiated the development of a commercially successful industry was doubtful.

In 1905 the decision to undertake an eradication rather than a suppression or biological control program may have been determined by the general philosophy of that time. The following quote from R. H. Shamblin (1924), superintendent of the United States date palm garden, shows this bias in favor of eradication: "I am still fully convinced if our palms were sprayed regularly at least four times a year we'd have better dates and less date pests. Personally, I believe in eradication rather than control." Although the decision to eradicate may have been hastily made and biased, it might also have been correct for this particular insect. The date palm scale has many characteristics that make it a favorable target for eradication. It is native to North Africa and the Middle East, so natural barriers help create an effective quarantine. Quarantine is easily established by restricting entry of palm shoots from the Old World or treating them as they arrive. The scale also has few hosts, low fecundity, and little natural dispersal. The scale is easily detected on the leaves with visual inspection, and although scales hidden under the leaf are not immediately visible, they are in low numbers and will become evident later (Boyden 1941). Thus, survey crews can easily map out infested areas. This process is helped by the limited geographic range of host plants that restrict the possible infested area. All that remains to complete a thorough eradication program is to find a suitable method to kill the pest on the imported offshoots and in the infested orchards.

Control efforts were first directed towards the imported offshoots. These were treated with distillate sprays, resin wash, repeated kerosene application, soap emulsions, and/or hydrogen cyanide fumigation (ibid; Forbes 1907). These were considered the most effective tools available, and supervisors were quite confident they would achieve control, but this was not the case. Palms in a number of orchards in Arizona and California were found to be infested. Application of kerosene and cyanide fumigation continued, often killing the plant, but not always controlling the date palm scale. The scale was spreading.

In 1905 R. H. Forbes of the Arizona Agriculture Experiment Station began an investigation. He discovered the scale was protected underneath the leaf bases. Soon afterward, he noticed that many of the palms severely burned in the 1906 San Francisco fire had recovered (Cockerell 1907). Using these two observations he began experimenting with a gasoline torch and with cropping back the old leaf bases. With this method the protected scales were reached. The tree was not killed by the torching, but the fruit production was lost for at least two and a half years. What could be called a "slash and burn" technique was adopted as the desired control method. It is likely that fumigation or spray would have worked equally well after defoliation and there may have been less crop loss. At the time the decision to burn back the foliage was made because of convenience and reliability.

The first official act of quarantine and inspection came with the appropriation bill of the Arizona Agricultural Experimental Station passed by the Territorial Legislature of Arizona in March 1907 (Boyden 1941). This bill authorized state officials to inspect date palms brought into the territory for insect pests. In 1909 the territorial legislature passed an additional act that authorized and regulated plant entry, inspection and pest control, and the enforcement (by state entomologists) of quarantine regulations. On March 24, 1913, a Federal Notice of Quarantine (No. 6) became effective under the Plant Quarantine Act. From this point forward, the control attempts moved from disconnected control efforts by individual groups to a multistate program.

Eradication efforts officially began in 1914 using the available tools: quarantine, treatment (defoliation and torch), and the removal of wild seedlings or noncommercial palms. The four centers of infestation were the Salt River Valley in Arizona, Laredo, Texas, and the Imperial and Coachella Valleys in California.

In Texas, treatments from 1915 to 1919 removed the scale. Unfortunately, in Arizona and California eradication was more difficult. Efforts were concentrated in those commercial orchards that were initially infested. Here they were able to contain and often remove the scale but unable to stop its reintroduction and spread from wild seedlings and abandoned plantations. For example, in the Tempe garden and city of Yuma, Arizona, there were hundreds of infested palms in 1919. By 1917, no infested palms were found in the experimental garden, and only five were found in Yuma. But the door-yard plantings and unattended palms outside the control area continued to harbor the scale.

Many of these initial attempts were also carried out half-heartedly because growers did not want to damage their young orchards with the harsh control methods. In El Centro, California, the scale continued to advance from 1915 to 1923, until a systematic inspection and thorough treatment program was begun. In the Coachella Valley the situation was worse because of the large number of plantations and wild seedlings. Boyden (1941) reports that many plantations once free of the scale were quickly reinfested.

There were attempts to control these reservoir populations. In 1923, a seedling plantation with about twenty-five hundred infested palms was destroyed, except for twenty-three palms that were treated. But the scale continued to spread. The reason was a failure to adhere to strict requirements for eradication. The survey crews were small and spent most of their time on large commercial plantations, leaving wild seedlings that acted as reservoir hosts. Also, inspection and control methods may have been relaxed because of the difficulty in reaching each palm and the desire to avoid harsh treatment of the valuable date palms.

In 1927 the increase in infested groves and the realization that control efforts were inadequate led to the reorganization of the eradication program. To meet this need, the Federal Horticultural Board appropriated an additional $25,000 in December 1927. B. L. Boyden was appointed to head an intensified effort to deal with the problem.

A survey of the infested regions was begun to determine whether eradication was feasible. With the additional funds, the inspection force was increased, and every planting was systematically recorded on a map along with such information as owner, presence or absence of scale, type of planting, and tree age. In about one-half year, 220,000 plants were found, of which 69 percent were seedlings. One percent of all palms were infested, and nearly all of those infested were seedlings. All plants were within the four areas, a total of 1,000 square miles.

In September 1928, after the preliminary survey was finished, date growers met with representatives from the Arizona Office of Entomology, the California Department of Agriculture, and the Federal Horticulture Board to determine control procedures (Boyden 1929). They considered the date palm scale to be a serious pest and believed that to seasonally control it would be both difficult and expensive. Biological control was mentioned but quickly dismissed as too uncertain. The meeting participants decided that eradication was feasible because of previously mentioned factors. At this meeting it was stated that eradication meant the elimination of every single scale.

The previous efforts at eradication had been inadequate primarily because of insufficient funds to deal with the abandoned seedlings and the lack of necessary organization to be completely systematic. The 1928 survey showed that there were far more of these neglected plants than standard ones and that some of these were infested and thus a reservoir from which the pest could spread to other areas. More funding was sought and received, an amount totaling $234,160 (most of which came from the federal government, with California contributing $50,000 and Arizona $20,000).

The eradication project was a cooperative effort between Arizona, California, and the federal government. The headquarters were set at Indio, California, in the Coachella Valley. The eradication program now consisted of three operations: survey, to locate the hosts; inspection, to discover infested palms; and treatment, to eliminate the scale.

The most important change was the increased intensity of the survey and inspection crews. The 1928 survey was rather hurried, and there was only a maximum of twelve inspectors. Because many palms had been missed, a completely thorough survey had to be undertaken to find all date palms as well as Canary Island, doum, and fan palms. New inspectors were trained. By the end of 1929 there were thirty-six full-time inspectors. The inspectors rotated districts, so that each person stayed in one district only a short while, thus increasing the chance that another person would find trees missed by previous inspectors. The survey crews were classified as *routine* when work at one location lasted five or more days (usually large, commercial orchards) and *scout* if the inspection lasted four days or less. Scout inspections became important in locating new infestations on isolated hosts.

Inspecting the trees for scale was often difficult because many of the abandoned seedling plantations were vastly overgrown, making location and counting of the palms laborious. These areas had to be cleared, and many abandoned palms were dug up (if the owner permitted it). From 1929 to the close of the project, 47,881 date palms were dug out and destroyed. Inspecting the tall trees was also a challenge. A platform on a truck was used, but extension ladders were also needed to get the very tallest trees. Each tree had to be inspected several times, the number of inspections depending on the degree of infestation and the area. Wherever infestations were found they were treated by defoliation and torching as in the previous effort (Boyden 1941). Besides these efforts, the quarantine of offshoots was maintained.

The annual progress of the program is summarized in table 14.1. The number of infested plants quickly declined, and by 1935 none were found. The project continued for another half year, during which time no date

Table 14.1   Number of Properties and Palms Found Infested with the Date Palm
Scale, *Parlatoria blanchardi,* 1928–1936 (from Boyden 1941)

Year	Total Infested Properties	New Infested Properties	Infested Palms
1928	80	45	3,176
1929	95	54	804
1930	54	17	389
1931	15	2	43
1932	7	2	44
1933	4	0	10
1934	1	0	3
1935	0	0	0
1936	0	0	0

palm scales were found. The project ended June 30, 1936. Since this time no
date palm scale has been found on date palms in the United States (Ray
Gill, personal communication, October 1988).

Evaluation

Eradication of the date palm scale was successful and represents one of the
few cases in which an insect was completely eradicated from a relatively
wide region. Several lessons can be learned from this experience. Although
the date palm scale was seemingly an ideal candidate for successful eradication, it took an extensive effort to accomplish this. The first attempt failed
because the methodology, though good, was not adequately carried out.
Successful eradication required a large, well-organized program that would
systematically and thoroughly cover an area, find every last host, carefully
treat, and keep accurate records so that the progress could be monitored and
treatments repeated. To do this required the cooperation of government,
growers, and citizens within the area. One could well imagine that people
might balk at having their trees defoliated and torched or completely dug
out. There was no record of any major complaints in this case, and the
coordination of the various groups seemed to work smoothly. These factors
surely played an important part in the success of the program.

Before an eradication program can be seriously considered, the true costs
and benefits should be examined. In this case, the costs apparently were not
completely weighed. Each tree treated lost two years of its production and
may not have been fully productive for a longer period. This cost was not
even considered. Another cost was a reduction in the growth rate of the
industry. It was very difficult and expensive to start new plantations in areas

outside of the eradication zone because of the quarantine on movement of offshoots. A method was finally developed to clean the offshoots by heating them at a precise temperature. Although the process was expensive, it removed the scale without damaging the valuable seedling (Swingle 1927). And, of course, there was the direct cost of the eradication program itself. This was $234,160 for the years 1929–36. This was almost equal to the value of the crop grown during this time in California. In 1980 dollars this comes out to be $1,400,000 (California Coop Reporting Service 1940).

One problem with eradication is that the pest can always be reintroduced, a problem that should be a consideration before an eradication program is attempted. Since offshoots of date palms are not allowed into this country from foreign areas, quarantine of *P. blanchardi* is effective and easily controlled. But it is of interest that from 1968 to 1972 forty-nine of these scales were intercepted at the borders, indicating that reintroduction is quite possible (USDA 1957–72).

Finally, to assess the value of the eradication one must compare the costs and benefits to possible alternatives. Biological control of the date palm scale has been successful in many areas similar to the date-growing regions of the United States and could have been an alternative. It may also have been less costly, not only in the direct cost but in the cost to the industry: the loss of production and the retardation of industry growth. In the 1928 meeting in which eradication was evaluated, however, biological control received only passing notice and was quickly rejected as being uncertain. Since eradication had failed previously, the decision to attempt eradication a second time seemed to reflect a bias towards eradication rather than a careful weighing of the costs and benefits. With hindsight, eradication appears to have been worthwhile: the pest was removed, and costs since eradication have been low. Whether alternative control measures would have worked as well or have been less costly will not be known.

Literature Cited

Bartlett, B. R., et al. 1978. Introduced parasites and predators of arthropod pests and weeds: A world review. USDA *Agric. Handbk* 480, 113–14.

Boyden, B. L. 1929. Progress of date scale eradication campaign. *Report Sixth Annual Date Growers' Institute,* 13–14.

————. 1941. Eradication of the Parlatoria date scale in the United States. *U.S. Dept. Agric. Misc. Pub.* 433.

Butan, D. 1975. Insect pests of fruit crops and their control. *Pesticides* 9:40–42.

California Coop Reporting Service. 1940. California production, average price to grower, and farm value estimates of avocados, dates, persimmons, and pomegranates. April 8.

Cockerell, T. D. A. 1907. The scale insects of the date palm. Univ. of Ariz., *Agric. Exp. Sta. Bull.* 56.

Commonwealth Institute of Entomology. 1962. *Distribution maps of pests.* Map 148.

Forbes, R. H. 1907. The extermination of date-palm scales. Univ. of Ariz., *Agric. Exp. Sta. Bull.* 56.

New York Times. 1928. April 29, sec. 5, 18.

Nixon, R. W. 1959. Growing dates in the United States. USDA-ARS *Ag. Info. Bull.* 207.

————. 1971. Early history of the date industry in the U.S. *Report of the Forty-Eighth Annual Date Growers' Institute* 48:26–30.

Popenoe, P. 1973. The Date Palm, ed. Henry Field. Miami, Fla.: Field Research Projects.

Shamblin, A. J. 1924. Eradication and control of the date scale. *Report of the First Date Growers' Institute,* February 29 to March 1.

Siddig, A. S. 1975. Field control of the scale insect *Parlatoria blanchardii* Targ. (Diaspididae) infesting date palm in Sudan. *J. Hort. Sci.* 50:13–19.

Smirnoff, W. A. 1957. La cochenille du palmier dattier (*Parlatoria blanchardi* Targ.) en Africa du Nordi, comportement, importance, economique, predateurs, en lutte biologique. *Entomophaga,* March.

Stickney, F. S. 1934. The external anatomy of the Parlatoria date scale, *Parlatoria blanchardi* Targioni-Tozzetti, with studies of the head, skeleton and associated parts. *U.S. Dept. Agric. Tech. Bull.* 421.

Swingle, W. T. 1927. Date plantings free from pests begun in irrigated southwest. *U.S. Dept. Agric. Yearbk,* 274–76.

U.S. Department of Agriculture (USDA). 1957–72. Plant quarantine division list of intercepted plant pests. Washington, D.C.: USDA-ARS.

15

Gypsy Moth in the Northeast and Great Lakes States

Steve H. Dreistadt and Donald C. Weber

In 1868 the gypsy moth escaped from the Medford, Massachusetts, laboratory of Leopold Trouvelot, who had imported the insect from Europe to
evaluate its potential for silk production. Twenty years after this initial
escape, the public outcry against hordes of gypsy moth larvae prompted
intensive efforts to rid the Commonwealth of Massachusetts of this pest.
The first insect targeted by organized eradication efforts in the United
States, the gypsy moth, *Lymantria dispar* (L.) (Lepidoptera: Lymantriidae), has continued to spread and is now the most notorious hardwood forest pest in the nation.

No forest insect in the United States has been the focus of more research
and control than the gypsy moth. Eradication has been abandoned in the
generally infested northeastern United States, where gypsy moth management efforts now focus on suppressing outbreak populations that reach
levels high enough to cause defoliation. But many states *outside* the general
infestation area have continued to pursue eradication by treating low-level
innocuous infestations that are causing no defoliation and would go undetected for at least a few years without the use of sensitive pheromone-
baited traps. This chapter appraises both the early eradication attempts in
the Northeast and later efforts in the Great Lakes region.

Life History, Damage, and Ecology

The male gypsy moth is approximately one inch long with a grayish dorsal
anterior and distinct wavy black bands on its dark brown wings. Females are
slightly larger and mostly white with dark, inverted V-like wing bands.
Males are attracted in summer to the pheromone of the female. A gravid

Drs. R. T. Cardé, J. S. Elkinton, and C. P. Schwalbe provided helpful suggestions and
criticisms in their reviews of this manuscript.

female typically lays one cluster of several hundred eggs, which she protec-
tively covers with setae from her abdomen. This univoltine species survives
fall and winter as diapausing fully-formed larvae within the egg. Egg hatch
coincides with leaf flush and warming spring weather.

Natural gypsy moth dispersal occurs primarily among first instar larvae,
particularly at high population densities. Larvae balloon down from the
canopy on silken threads and with the aid of their long setae sail off in the
wind. Long distance transport occurs among eggs and sometimes pupae
that are attached to nursery stock, outdoor furniture, and vehicles traveling
through or moving from generally infested areas.

Larval development requires approximately six weeks. Males typically
have five instars and North American females six. Each stadium lasts four to
ten days. Metamorphosis within the 3/4 to 1-1/4 inch teardrop-shaped pupa
requires about two weeks.

High-level gypsy moth populations cause temporary tree defoliation and
some mortality. Property values, wildlife, recreation, and watershed may be
temporarily affected (White and Schneeberger 1981).

Deciduous oaks, birch, larch, aspen, willows, and some fruit trees are
primary gypsy moth hosts (Campbell 1979; Hough and Pimentel 1978;
Houston 1979). Defoliation-induced tree mortality may range from none to
more than 50 percent in repeatedly defoliated stands (Campbell 1979;
Brown, Halliwell, and Gould 1979). Over a ten year period in Connecticut,
tree mortality increased from one percent per year in undefoliated stands to
an average annual mortality of 3.5 percent in stands subject to two successive
defoliations. From this Stephens (1981) concluded that, "Repeated defoliation will accelerate loss of oaks and increase species less susceptible to
defoliation, but the forests will not be destroyed."

Temporary defoliation and the loss of some trees are often of less public
concern than the insect nuisance. The sometimes high numbers of caterpillars or fluttering moths can annoy and even frighten people. There is also
the problem of skin rashes, or in some persons more severe reactions, from
contact with the caterpillars (Etkind et al. 1982; Shama et al. 1982).

Probably the most obvious way to estimate the impact of the gypsy moth
is to look at the money spent in control and eradication efforts. More
pounds of pesticides have been applied against the gypsy moth than for any
other forest pest in North America, with the possible exception of the
spruce budworm, *Choristoneura fumiferana* (Clemens) (National Research
Council 1975). From 1905–40, Massachusetts alone reportedly spent $25
million on gypsy moth control (Dunlap 1980). Gypsy moth control costs by
the 1970s were placed at $100 million (Daly, Doyen, and Ehrlich 1978), and
the cumulative nationwide costs by the early 1980s probably exceeded $200
million.

Gypsy moth populations in the northeast are most often innocuous but periodically reach levels of extreme local or regional abundance. During 1981, a record 12.9 million acres experienced light to heavy defoliation (fig. 15.1). By 1984, noticeable defoliation had dropped tenfold to below 1 percent of the approximately 120 million forested acres in the fourteen states now considered to be within the general infestation (USDA 1985).

Despite decades of research, no sweeping conclusions can be drawn as to why gypsy moth sporadically reaches levels of extraordinary abundance. Drought, low or highly fluctuating temperatures, foliage quality, and natural enemies and disease are all influential in determining gypsy moth abundance (Wallner and McManus 1981). Selective cutting of conifers, reforestation of poor piedmont soils abandoned after agricultural "mining" of soil nutrients, and the introduction of *Endothia parasitica* (Murr.), which blighted the formerly dominant and gypsy-moth-resistant American chestnuts, have increased eastern hardwood forest susceptibility to this pest (Smith 1976; Stephens 1976).

Under some circumstances, both established and newly introduced gypsy

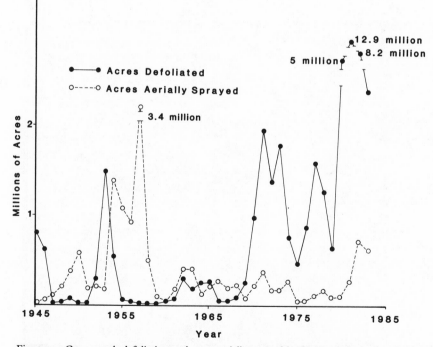

Figure 15.1 Gypsy moth defoliation and acres aerially sprayed in the Northeast and Michigan, 1945–1983 (Sources: Hanna 1982; Moore 1983; National Research Council 1975; USDA 1985; USDA, undated, Gypsy moth suppression acreage; Wolfe 1985).

moth populations may long remain at extremely low levels or become naturally extinct without any human intervention. During 1981–84, apparent gypsy moth infestations were identified by moth captures during three consecutive years in Berkeley, Mill Valley, and Santa Cruz, California. Because backyard searches for egg masses were unsuccessful in precisely pinpointing the populations, no eradication projects were initiated, and no further moths have been detected at these sites (Henry 1981; California Department of Food and Agriculture [CDFA], undated, Gypsy moths detected in 1982; CDFA, undated, Summary–1983 gypsy moth finds; Siddiqui 1984; CDFA 1984).

Gypsy moth is rare and may have become extinct in the British Isles. No defoliation has ever been reported in some portions of the insect's worldwide range (Giese and Schneider 1979) (see fig. 15.2 for approximate gypsy moth distribution). Gypsy moth populations in Michigan remained innocuous for thirty years (Dreistadt 1983a). Gypsy moth at some sites in the northeast maintained relatively stable in low-level, nondefoliating populations over several decades (Campbell, Hubbard, and Sloan 1975a, 1975b; Campbell 1976; Campbell and Sloan 1976).

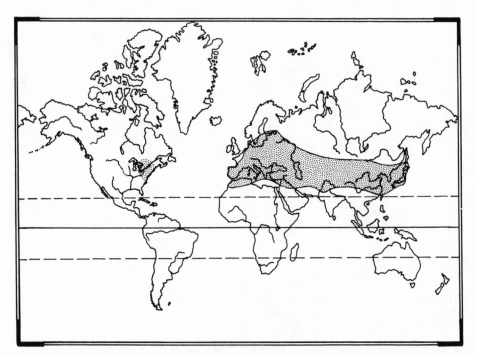

Figure 15.2 Distribution of the gypsy moth, *Lymantria dispar* (adapted from Commonwealth Institute of Entomology, 1981).

Control

Several methods have been used in eradication campaigns against gypsy moth. Early attempts in the Northeast combined primarily manual methods with some chemical and cultural controls. Hand removal of egg masses was laborious, time-consuming, and imperfect, even when combined with filling-in of tree hollows. Burlap-banding of trunks, necessitating daily checking for hidden larvae, had similar drawbacks. Cultural control by elimination of gypsy moth hosts was employed in the first eradication attempt by completely clearing and burning the infested forests (Forbush and Fernald 1896).

Chemical control has been the most widely used method for attempted eradication. Historically, lead arsenate and DDT were the major chemicals employed. Four synthetic chemicals are now most commonly used against *L. dispar:* carbaryl, trichlorfon, acephate, and diflubenzuron. But at least thirty other insecticidal ingredients are present in different products registered for gypsy moth control (Rachel Carson Council 1987).

Carbaryl (Sevin) is a broad-spectrum contact carbamate insecticide, typically applied at 1.12 kg/ha active ingredient (1 lb/acre AI). The persistence is a few days to a few months, up to at least eighty days on leaves (Neher, Segawa, and Oshima 1982). It is more toxic to Hymenoptera than to the target insect, somewhat less toxic to Diptera (White, McLane, and Schneeberger 1981). Carbaryl is teratogenic in beagle dogs and possibly in guinea pigs and rats (USEPA 1980) and has a variety of detrimental effects on nontarget organisms (Weis and Weis 1979; Brown 1978; Pimentel 1971). Carbaryl has been used in numerous eradication attempts in the eastern and western United States since 1958.

Trichlorfon (Dylox) is an organophosphate that has been widely applied aerially as an oil dispersion at 1.12 kg/ha (1 lb/acre) (White, McLane, and Schneeberger 1981). Pennsylvania used trichlorfon for several years as its major suppression tool, ending its use in 1983. It has rarely been used in eradication attempts to date (Pennsylvania Governor's Press Office, press release, September 9, 1982).

Acephate (Orthene), another organophosphate contact toxin, is typically applied as a water solution at one-half the rate of trichlorfon. According to White, McLane, and Schneeberger (1981), "population reduction, based on egg-mass surveys, was not dramatic. . . . Acephate does not offer many more advantages than either carbaryl or trichlorfon." It has a relatively high contact toxicity to the pupal parasite *Brachymeria intermedia* (Nees), but not when already on foliage. The tachinid *Compsilura bifasciata* How. is relatively resistant (ibid; Respicio and Forgash 1977). Acephate is very soluble in water, relatively nonpersistent, and is readily translocated from soil to leaves (White, McLane, and Schneeberger 1981). Acephate has enjoyed scattered use against gypsy moth, including small-scale eradication attempts, particularly in Washington from 1979 through the early 1980s.

Diflubenzuron (Dimilin) produces mortality by inhibiting ecdysis in many arthropods, including the gypsy moth. Shortly after the Environmental Protection Agency (EPA) registered it in 1976, California twice sprayed it over 2,900 acres in San Jose in a gypsy moth eradication campaign that included carbaryl applications from the ground (CDFA 1977, 1978). In April 1979, Dimilin use was limited to cotton and against gypsy moth, in uninhabited and sparsely inhabited areas three miles from "bodies of water." Chronic health effects of the active ingredient and two metabolites in mammals prompted these restrictions (Troast and Sanders 1979; F. Gee, personal conversation with D. C. Weber, 1981). These restrictions were relaxed in 1985 to allow application in urban and residential areas and to within one mile of water (Uniroyal Chemical, Dimilin 25W Insecticide label and attachments, 1985). Application rates are quite low: 0.5–1 oz/acre (0.034–0.067 kg/ha). Dimilin has a variable, often detrimental, effect on parasites (White, McLane, and Schneeberger 1981). Its persistence is longer than other gypsy moth insecticides. As much as 90 percent of the active ingredient remains on or in the leaves after one to two months and 16 percent remains in the soil after one year. Dimilin is a hazard to aquatic arthropods and has some capacity for bioaccumulation in aquatic habitats (ibid; Troast and Sanders 1979).

All four commonly-used chemicals are broad insecticides. This has implications for eradication programs. If natural enemies of gypsy moth are present, insecticide treatments increase the risk associated with a failed eradication attempt: survivor moths may be released from these controls. Other potential pest species in the system may be released from their natural enemies as well.

Eighty-six pathogens—bacteria, protozoa, fungi, and viruses—have been isolated from natural gypsy moth populations (Gerardi and Grimm 1979). Two pathogens, *Bacillus thuringiensis* Berliner (*B.t.*) and a nucleopolyhedrosis virus (NPV), are used in control.

The pathogen *B.t.* (Dipel, Thuricide, Biotrol) is a stomach poison to many Lepidoptera but is nontoxic to gypsy moth parasites (Dubois 1981) and vertebrates, including humans. It is relatively nonpersistent on foliage, but increased application rates and improved adjuvants have led to efficacy equivalent to Dimilin (Jobin 1982; Miller and West 1987) and Dylox (PDER 1982). The pathogen was first used in eradication efforts in McHenry, Illinois (1980–81), and in combination with carbaryl in Santa Barbara, California, in 1981. Both eradication attempts were apparently successful (Dreistadt 1983b). Many states now use it as their primary insecticide in eradication projects, and it has also been used in state-administered suppression programs in Pennsylvania, Maryland, and other eastern states (Pennsylvania Governor's Press Office, press release, September 9, 1982; W. R. Hickman, letter to D. C. Weber, February 18, 1983; NYDEC 1983; MSDA 1983; MDEM 1982).

The pathogen NPV, also known as wilt, made its first appearance in North America and Europe in 1907 and by 1915 had spread over the entire range of the gypsy moth, probably by accidental introduction (Glaser 1915). The virus causes population collapse when high densities of larvae are present (Lewis and Yendol 1981). It is a stomach poison, specific to the Lymantriidae. As a commercial product, NPV was registered as Gypchek by the U.S. Forest Service in 1978. Its use is not widespread because of relatively high cost and the necessity for two applications. Only one project (in Wisconsin) employed NPV in attempted eradication.

Pheromones have played a prominent role in initiation and evaluation of modern eradication attempts. Gyplure, widely used for detection purposes for about ten years starting in the early 1960s, was subsequently shown to be inactive and incorrectly identified as a pheromone (Jacobson, Schwarz, and Waters 1970). Conclusions that no gypsy moths were present when none were detected by gyplure, which were reached most notably in Michigan, were therefore completely invalid (Dreistadt 1983a). Disparlure replaced gyplure in 1972. A few years later the (+) enantiomer was found to be about ten times as potent as the racemic mixture in use; by 1980 its use was widespread. Improved provision for steady volatilization of the active ingredient has made traps effective over the course of several months (Elkinton and Cardé 1981). But typical trap densities used in extensive trapping programs are likely not to immediately detect low-level infestations (Knipling 1979), and delineation of a proposed eradication area requires additional intensive localized trapping.

Disparlure use as a suppressive technique is in the experimental stage. Isolated infestations in Appleton, Wisconsin (WDA 1982), and in Monona, Wisconsin (K. P. Robert, letter to S. H. Dreistadt, 1985), have apparently been eradicated using three traps per acre as the sole suppressive measure. Such high-density trapping is likely to fail unless scrupulously coordinated, and it is useless at high moth densities. In 1983 Virginia used a 4,000-acre aerial application of racemic disparlure, formulated in Hercon Disrupt flakes, to combat an isolated infestation in Floyd County. No moths were caught with intensive trapping in the following two mating seasons (C. L. Morris, conversation with D. C. Weber, 1985).

Inundation of sparse, isolated gypsy moth populations with sterile males is a method receiving increased emphasis among USDA-sponsored eradication programs. Three versions of this technique have been employed in eradication attempts:

1. fully sterile males introduced as pupae (Berrien County, Mich., 1980–82);
2. substerile males introduced as pupae, resulting in sterile F_1 offspring (Horry County, S.C., 1982);

3. introduction of F_1 sterile moths as egg masses (Darke County, Ohio, 1984; Bellingham, Wash., 1985–86; Manor, Wash. 1986–87).

All aim to produce an overflooding ratio (typically 100 : 1) of sterile to fertile males, to assure cessation of reproduction. Based on two years of negative catches (32 traps/mi^2) for both the Michigan and South Carolina sites, the populations are considered eradicated (Mastro 1984; Mastro and ODell 1985). Release of sterile eggs is less costly, since egg masses may be stockpiled from lab colonies and placed out in the field only once per season. The third technique is therefore the preferred approach, should field trials prove successful (Mastro et al. 1985; Mastro, telephone conversation with D. C. Weber, 1987).

Introduction of natural enemies of the gypsy moth represents the longest and most extensive effort at classical biological control in the United States. During the periods 1905–14, 1922–33, and since 1963, more than forty parasitoid species and several beetle predators have been imported. Insectary culturing before release has produced millions of organisms. Ten parasitoids and two predators have established (Hoy 1976; Nichols 1980; Smilovitz and Rhodes 1972). Opinions differ on the effectiveness of established parasites at different moth population levels (Reardon 1981). DeBach (1964) rated the parasite-release efforts a "partial success." Parasitoids have generally not been employed in eradication programs for gypsy moth, although at least one, the egg parasitoid *Ooencyrtus kuvanae* (Howard), has been introduced in Oregon, where eradication is being attempted over several hundred thousand acres, primarily through aerial applications of B.t. In combination with either B.t. or NPV treatments, parasite releases complement control by parasitizing the residual population and, in the case of NPV, vectoring the virus (Raimo and Reardon 1981).

For nearly a century, the evolving pest management strategies described above have been employed singly or in various combinations in the United States. Of special interest are the attempts to eliminate gypsy moth populations, particularly given the controversy surrounding eradication efforts in certain areas of the nation.

History of Eradication Efforts

Massachusetts, 1890–1900

From 1890 to 1900 Massachusetts spent about $1.2 million attempting to eradicate gypsy moth from an area of about 200 square miles (MBA 1902). This cost is equivalent to more than $16 million in 1983 dollars (USDC 1975, 1982). The most important control measures employed were: (1) manual removal and/or destruction of egg masses; (2) destruction of larvae using burlap tree-banding; (3) cutting and burning of vegetation; and (4) spraying of infested trees and shrubs with Paris green (copper acetoarsenite) and later lead arsenate. Spraying was confined to dense populations of caterpillars, amounting to less than 3 percent of those trees burlapped in every year except 1891. Manual efforts peaked in 1899, with more than 3 million trees burlapped (Forbush and Fernald 1896; MBA 1891–1900).

The elimination of the final surviving moths posed the most difficult biological and political barrier to the first eradication attempt. The ability of the moth to avoid visual detection at low levels was a major handicap in the absence of pheromone technology. The time-consuming nature of manual control methods aggravated the political situation. Funding was exclusively from the state, and legislators were concerned with defoliation and nuisance caused by the moth. Their sense of urgency bore "a direct relation to larval density" (Dunlap 1980). As the eradication efforts reduced moth numbers, their reluctance to provide more funding resulted in the termination of the program.

In the years following the abandonment of this first eradication attempt, the moth spread over all of New England. Influenced by the phenomenal success against cottony cushion scale in California, a massive parasite importation program began in 1905. Several important natural enemies, including NPV, were established. Two significant developments followed in the early 1920s: the establishment of a federal "barrier zone" along the Hudson River valley, north along the Vermont-New York border to Canada, and the discovery of a large infestation in northern New Jersey.

New Jersey, 1920–1932

In June 1920 the moth was discovered in a blue spruce plantation near Somerville. Traced to nursery shipments from Holland in 1911, the infestation was scattered over nearly 1,000 square miles when detected (Weiss 1921; Felt 1942). A twelve-year $2.5 million state-federal eradication effort was reportedly successful (NJDA 1932; Worrell 1960; Nichols 1961).

Methods were virtually identical to those employed in Massachusetts earlier, but with primary emphasis on lead arsenate spraying (Weiss 1921–23). There were 576 tons of lead arsenate dispersed at an average rate of 50 lbs/acre (Weiss 1921–27). For the first four years, about 40,000 trees were either burlapped or sticky-banded. The number of people employed in this attempt was comparable to the number in the Massachusetts campaign (NJDA 1932; Forbush and Fernald 1896; MBA 1896–1900).

The reported success in New Jersey may have arisen from: (1) immediate and sustained appropriations (half state, half federal); (2) knowledge of effective spraying techniques, with no need for research, experimentation, and fabrication; (3) ability to trace spruce shipments to satellite infestations; and (4) such other unknown factors as weather, genetics, and mortality from natural enemies. But the assertion of eradication in New Jersey should not be accepted uncritically. Although crude pheromone traps were distributed beginning in 1929, visual detection remained the principal method used. As late as 1934, "mop up" operations found 152 larvae and led to spraying of 150 acres (NJDA 1935). Within twenty years, the gypsy moth became recognized as a permanent resident of the Garden State.

Pennsylvania, 1932–1943

Following the establishment of the barrier zone, several infestations to the west and south were reportedly eradicated (Felt 1942; Nichols 1961). One large colony, discovered in Luzerne County, Pennsylvania, in 1932, proved intractable. Quarantine and eradication efforts commenced immediately. Large numbers of workers from the Works Progress Administration—a maximum of 1,100 in 1938—were needed to undertake labor-intensive measures identical to those employed in the Massachusetts and New Jersey campaigns. Ground spraying with lead arsenate was the chief technique. Creosote treatment of egg clusters and burlap banding were employed in areas inaccessible to spray rigs. Workers thinned and burned more than 10,000 acres of forest. The cost of spraying was high—about $25 per acre— and the thirty-five rigs could cover only 10,000 acres per season (Nichols 1961). By 1942, federal and state governments had spent more than $4.5 million, and nearly 1,000 square miles remained infested (Felt 1942). Because of the size of the area, lack of personnel, and equipment limitations, "hope of eventual eradication was all but abandoned" (Nichols 1961). Meanwhile, by 1939 the barrier zone was generally infested (Worrell 1960; McManus and McIntyre 1981).

DDT in the Northeast, 1956–1958

Sheals (1946) first made public the USDA plans for a new and massive eradication campaign using DDT. In 1952 the USDA gypsy moth study group recommended the entire infested area (then 30 million acres) "be treated for eradication in a period not to exceed five years." The immediate objective was to reestablish the barrier zone, eradicating the moth within the zone and to the west. Meanwhile, a strengthening of quarantine and federal assistance for chemical suppression within the generally-infested area would reduce the probability of spread (Perry 1955). The eradication, optimistic officials thought, could then sweep eastward. Representatives from the USDA met with state officials to decide on a regional course of action. Pennsylvania, New Jersey, and uninfested states supported an all-out eradication attempt; New York was ambivalent, and New England balked. As a compromise, eradication was adopted only as a "tentative ultimate goal" to be reassessed yearly, and the overall program was one of "eradication and/or control" (Worrell 1960). The USDA continued to consider the program one of eradication.

Confidence in DDT ran very high among economic entomologists: "The gypsy moth can be eliminated from any given area with a single application of one pound or less of DDT applied per acre as an oil solution" (Popham and Hall 1958); "Applications made during any state of larval development will result in complete eradication" (Nichols 1961). Airplanes revolutionized application. One large cargo plane "could treat in one hour an area that formerly took ten hydraulic units a whole season." The cost—about $1/acre—was dramatically lower (Nichols 1961).

In 1956, in a program that was two-thirds federally funded one million acres in New Jersey, Pennsylvania, and New York were aerially sprayed with one pound DDT in an oil solution per acre (Worrell 1960). The treatment next season of more than 3 million acres in the same three states created a clamor of public opposition, and "the first serious environmental litigation against a pest control program" (Dunlap 1980). Several objections were raised: (1) nuisance of oily scum; (2) contamination of dairy farms resulting in residues in milk; (3) burning of foliage of truck crops and flowers; (4) contamination of organic gardens; (5) bee mortality; (6) fish and wildlife kills; (7) human health hazard, including contamination of water supplies; (8) sloppy, often repeated applications (Dunlap 1980; Waller 1958; Sumnick 1958; Brown 1961; Carson 1962).

Mainly because of persistent DDT residues on dairy forage, New York suspended eradication efforts. The 500,000 acres treated in 1958 were exclusively in Pennsylvania. This was the last DDT used in the northeast states eradication attempt (McManus and McIntyre [1981]; these authors incorrectly state that *no* DDT was used against gypsy moth after this date). An "aggressive holding program" continued in New York (Popham and Hall 1958). In 1959 only 86,000 acres were treated, all in New York, with carbaryl at 1.25 lb/acre. By 1961, eradication was abandoned, and moths appeared in numbers on Long Island, in New Jersey, Pennsylvania, and upstate New York (Nichols 1961; Carson 1962).

The massive eradication effort in the mid-fifties was not as much of a response to public pressure as earlier eradication attempts had been. Instead, it was largely an outgrowth of the advent of DDT and aerial application, as well as the desire of economic entomologists to make full use of them (Dunlap 1980). The campaign proceeded with a poor scientific basis and with inattention to public input and reaction. These two faults led to its failure and the rapid erasure of any benefits obtained by spray treatments.

Three areas of scientific weakness characterized the DDT campaign: insufficient evidence that DDT was perfectly eradicative, overconfidence in existing detection methods, and ignorance and negligence with regard to the ecological and potential health effects of the pesticide. The first point is critical, since failure of eradication forfeited all potential benefits. The second point is closely related, in that inadequate detection methods were used to confirm results of experimental and operational treatments. The traps in use were less attractive than virgin females, and formulation was unacceptable (Nichols 1961; Beroza et al. 1974). Visual surveys, a notoriously difficult method for detecting sparse populations, were cited as evidence of zero moth numbers following treatments.

Apart from detection obstacles, the concept of eradicating at least 60 billion organisms spread over a huge area is a biological absurdity. (The estimate of 60 billion early-instar larvae in 1954 [moderate outbreak year] is based on 30 million infested acres [Perry 1955], with 500,000 of them defoliated [Brown 1961], using density estimates given by Campbell and Sloan [1978] with 50 percent mortality [high estimate, J. S. Elkinton, personal communication, 1988]). Also ignored was the capacity for development of DDT resistance, already widespread in other species (Brown and Pal 1971) and later discovered in gypsy moth by Tomlin and Forgash (1972).

Ignorance and negligence characterized the USDA statements concerning hazards to wildlife and man from DDT. According to Fish and Wildlife Service research (Cottam and Higgins 1946), applications of more than one-half pound would damage fish and crustaceans; this effect was downplayed (Nichols 1961). Systematic monitoring of ecological side-effects was absent. Assurances of the harmlessness of DDT were thus hollow. Later studies of DDT revealed the ubiquitous distribution, persistence, and bioaccumulation of a long-term threat to human health, wildlife, and entire ecosystems (Dunlap 1981).

The second major fault of the campaign, its inattention to the public, is explicable by two avenues. First, economic entomologists had never had such powerful tools. Huge eradication efforts seemed to offer huge benefits. Yet, because of their vast scale, such programs also conferred an obligation under a democratic government to public participation. In their political naiveté, government scientists and administrators focused mainly on operational questions, largely ignoring the political implications. The second explanation for inattention, and even scorn, regarding criticism of the program was the overall tenor of American society during this period. McCarthyism and the Cold War tended to suppress criticism in all walks of life. For government officials to treat dissenters as "un-American" was not uncommon. Only because the scientific problems were so glaring with the DDT moth eradication campaign did the public outcry manage to change the government plan for an all-out eradication.

Michigan, 1954–1984

Michigan is of great interest in the history of the gypsy moth because the pest was first discovered there after the procedure of aerial application of DDT had become available. A complaint regarding caterpillars from a home-owner prompted roadside visual surveys that identified an extensive gypsy moth infestation in south-central Michigan in 1954. Approximately 58,000 pounds of DDT were aerially applied to over 100 square miles. More gypsy moths were found, and from 1954 through 1962 approximately 365,000 acres in Michigan were aerially sprayed with DDT (Hanna 1982). Gypsy moth was reportedly eradicated by 1963 (Dreves 1982), when the bait in gypsy moth detection traps in Michigan was switched from pheromone extracted from females to synthetic gyplure.

No gypsy moths were again discovered until 1966 when caterpillars were reported in the area previously sprayed with DDT. This led to aerial carbaryl application over 12,000 acres in 1967 and the reported eradication of this "new" infestation (ibid).

In 1972, when disparlure replaced gyplure as a trap attractant, the first moths detected in Michigan since 1966 were trapped in three counties. Improved traps in 1973 captured 1,828 moths in twenty-one counties. Car-baryl and diflubenzuron treatments were initiated, and after three years of treatments the Michigan Department of Agriculture (1976) reported that "considerable progress has been made relative to the eradication of the gypsy moth from Michigan." Infestations discovered by 1977 were report-edly eradicated in the counties of Macomb, Washtenaw (USDA 1982), Mecosta, and Berrien (Dreves 1982).

By 1980, however, when most traps in Michigan employed a refined disparlure consisting only of the (+) enantiomer, gypsy moths were de-tected in thirty-seven counties. Even so, Michigan has maintained a policy of eradicating apparently isolated infestations found outside of the south-central part of the state. More than 40,000 acres in sixteen counties were treated for eradication from 1980–84 (MDA 1984; USDA 1985). Despite this effort, gypsy moth has spread into at least seventy-three of eighty-three counties in Michigan (Moore and Hanna 1984). Millions of acres through-out much of the state are now infested, and defoliation was first recorded beginning in 1979 (ibid).

Despite repeated claims of eradication since the 1960s, substantial biolog-ical evidence indicates that gypsy moth has resided continuously in Michi-gan for more than thirty years (Dreistadt 1983a). Roadside visual surveys and five thousand natural pheromone-baited traps per year in a state with more than nineteen million acres of forests were inadequate to delimit the initial infestation. Gyplure, used in traps during 1963–71 when Michigan was claimed to be without gypsy moth except for one infestation discovered by the visual report of caterpillars, was later found to be ineffective at trapping the gypsy moth (Jacobson, Schwarz, and Waters 1970). The nearly 2,000 moths found in 1973 by improved trapping in twenty-one counties were not likely recent reintroductions.

Despite the failure of eradication in the state, Michigan has been focusing the major portion of its gypsy moth program budget on attempts to eradicate apparently isolated infestations detected outside the south-central part of the state. It would be less costly and more environmentally sound to initiate treatments in Michigan only where gypsy moth populations reach high enough levels to cause defoliation (Morse and Simmons 1978).

Other Great Lake States

Ohio, 1973–1985. Gypsy moth was first reported in Ohio as early as 1914, when an unrecorded number of egg masses were destroyed (W. K. Roach, letter to S. H. Dreistadt, 1985). From 1973 through 1985, Ohio conducted thirty-eight eradication projects in thirty communities (table 15.1). All but one employed carbaryl (Sevin 4-Oil), and/or trichlorfon (Dylox 1.5 Oil). Sterile moth releases were conducted north of Greenville (Darke County) in west central Ohio in 1985.

Ohio is bordered by general infestations in Pennsylvania to the east and in Michigan to the north, and at least the northeastern portion of the state must now be considered part of the general infestation. In 1986, one or more moths were detected in sixty-six of the eighty-eight counties in Ohio, and 14,854 moths were trapped in Ashtabula, the most northeastern country of Ohio.

Illinois, 1976–1985. From 1976 through 1985 gypsy moth eradication projects were conducted in 19 Illinois communities (table 15.2). The first gypsy moth eradication project in Illinois was conducted in Palos Park in 1976. This program employed two aerial applications of carbaryl (Sevin 4-oil), a procedure preferred by the Illinois Department of Agriculture (IDA) in any community that did not actively challenge the use of broad-spectrum insecticides. The first such challenge was in McHenry, fifty miles northwest of Chicago. In 1979, after 165 moths were trapped there, the IDA and USDA proposed aerial applications of carbaryl (Ginsburg 1981). Residents obtained a court injunction against the applications, and in 1980 the Concerned Citizens for the Environment and McHenry County Defenders paid $700 for ground spraying of infested properties using B.t. Three property owners denied access for the B.t. spraying, one of whom applied carbaryl to her property using a garden hose sprayer (M. Teta, undated).

Residents banded three hundred trees with burlap to monitor for caterpillars, and the USDA supplied two thousand pheromone traps, which vol-

Table 15.1 Gypsy Moth Eradication in Ohio (1972–1985)

City	County	Year	Acres	Method	Moths Caught
N. Ridgeville	Lorain	1972	—	—	7*
		1973	300	trichlorfon(1)^A	0
		1973	1	carbaryl(1)^G	—
Dayton	Montgomery	1974	—	—	4
		1975	295	trichlorfon(1)^{A,G}	0
		1975	15	carbaryl(2)^G	—
Kettering	Montgomery	1978	—	—	7
		1979	—	—	170*
		1980	320	trichlorfon(2)^A	9
		1980	15	carbaryl(2)^G	—
		1981	219	trichlorfon(2)^A	7
		1982	—	—	0
Catawba Island	Ottawa	1979	—	—	25*
		1980	285	trichlorfon(2)^A	137*
		1980	10	carbaryl(2)^G	—
		1981	600	trichlorfon(2)^A	13†
		1982	38	carbaryl(2)^A	0
Uniontown	Stark	1980	—	—	30*
		1981	237	carbaryl(2)^A	4
		1982	58	carbaryl(2)^A	0
Woodsong Pass	Hamilton	1980	—	—	5
		1981	15	carbaryl(1)^G	0
Ottawa Hills	Lucas	1981	—	—	14
		1982	160	carbaryl(2)^A	0
Bay Village	Cuyahoga	1981	—	—	14*
		1982	38	carbaryl(2)^{A,G}	0
Pepper Pike	Cuyahoga	1981	—	—	6
		1982	26	carbaryl(2)^{A,G}	0
Solon	Cuyahoga	1981	—	—	29*
		1982	58	carbaryl(2)^{A,G}	0
Salem	Columbiana	1981	—	—	5
		1982	38	carbaryl(2)^A	2
Newcomers-town	Tuscarawas	1981	—	—	7
		1982	93	carbaryl(2)^A	0
Barkcamp State Park	Belmont	1981	—	—	11
		1982	120	carbaryl(2)^A	0
Jefferson Lake State Park	Jefferson	1981	—	—	19
		1982	106	trichlorfon(2)^A	0
Tanager Woods	Hamilton	1981	—	—	12
		1982	75	carbaryl(2)^A	0
Paul Meadows	Hamilton	1981	—	—	7*
		1982	15	carbaryl(2)^A	0
Hopper Road	Hamilton	1981	—	—	7*
		1982	92	carbaryl(2)^{A,G ·}	0

(*continued*)

Table 15.1 (*Continued*)

City	County	Year	Acres	Method	Moths Caught
Mt. Vernon	Knox	1981	—	—	18
		1982	30	carbaryl(2)^A	27
		1983	93	carbaryl(2)^A	1
		1984	—	mass trapping	0
Anderson Township	Hamilton	1981	—	—	10
		1982	10	carbaryl(2)^A	7
		1983	58	carbaryl(2)^A	0
W. B. Turnpike	Portage	1981	—	—	27
		1982	35	carbaryl(^{2})^{A,G}	12
Blacklick	Franklin	1981	—	—	17
		1982	—	—	21
		1983	128	carbaryl(2)^A	0
Jefferson Township	Franklin	1981	—	—	1
		1982	—	—	17*
		1983	96	carbaryl(2)^A	0
Hyde Park area, Cincinnati	Hamilton	1981	—	—	8
		1982	—	—	23
		1983	102	carbaryl(2)^A	0
Sylvania Township	Lucas	1982	—	—	20
		1983	128	carbaryl(2)^A	35
		1984	90	carbaryl(2)^A	0
Little Turtle	Franklin	1983	—	—	20
		1984	50	carbaryl(2)^A	0
Zoar	Tuscarawas	1984	—	—	25
		1985	600	carbaryl(2)^A	1
Gates Mills	Cuyahoga	1984	—	—	18
		1985	90	carbaryl(2)^A	0
Findlay	Hancock	1984	—	—	18
		1985	80	carbaryl(2)^A	0
Sylvania	Lucas	1984	—	—	13*
		1985	30	carbaryl(2)^A	1
Greenville	Darke	1984	—	—	100*
		1985	20	sterile moths	?

Sources: W. K. Roach, specialist in charge, Plant Pest Control, Ohio Dept. Agric., letter to S. H. Dreistadt, Dec. 5, 1985; USDA 1985; Ohio Dept. Agric., Summary of site-specific environmental assessment activities for proposed gypsy moth control in Ohio, 1983; H. L. Ford, deputy director, USDA-PPQ, letter to W. Morris, director, plant health division, Agriculture Canada, 1985.

Notes: * = Other life stages found; † = Moths outside treatment area; (Number of applications); ^{A} = aerial application; ^{G} = ground application.

Table 15.2 Gypsy Moth Eradication in Illinois (1975–1985)

City	County	Year	Acres	Method	Moths Caught
			Treatment		
Palos Park	Cook	1975	—	—	190*
		1976	1,000	carbaryl(2)A	0
McHenry	McHenry	1979	—	—	166*
		1980	40	B.t.(2)G	—
		1980	—	mass trapping	72
		1981	100	B.t.(2)A	—
		1981	—	mass trapping	0
		1982	—	—	3†
		1983	—	—	0
Kildeer	Lake	1979	—	—	46
		1980	30	carbaryl(2)A	0
		1981	—	—	10
		1982	—	—	21
		1983	—	—	12
		1984	—	—	—
		1985	27	mass trapping	?
Lincolnshire	Lake	1980	—	—	45*
		1981	35	carbaryl(2)A	16†
		1982	—	—	3
		1983	—	—	6
Wheaton	Du Page	1980	—	—	69*
		1981	50	carbaryl(2)A	286*†
		1982	300	B.t.(2)A	—
		1982	300	mass trapping	103
		1983	230	B.t.(2)A	—
		1983	800	mass trapping	5†
		1984	—	—	1†
Diamond Lake	Lake	1980	—	—	17*
		1981	3	carbaryl(2)G	1†
		1982–83	—	—	0
Wood Dale/ Bensenville	Du Page	1981	—	—	1,479*
		1982	500	B.t.(2)A	—
		1982	—	mass trapping	166
		1983	989	B.t.(2)A	—
		1983	1,532	mass trapping	13
		1984	400	mass trapping	0
Napperville	Du Page	1981	—	—	36*
		1982	60	B.t.(2)A	17
		1983	40	B.t.(2)A	—
		1983	200	mass trapping	0
		1984	—	—	0
Crystal Lake	McHenry	1981	—	—	49
		1982	100	carbaryl(2)A	4†

(continued)

Table 15.2   (*Continued*)

City	County	Treatment Year	Acres	Method	Moths Caught
		1983	—	—	2
Lake Zurich	Lake	~~~	—	—	49*
		1982	80	carbaryl(1)^A	1
		1982	—	—	1
Lake Forest	Lake	,~~	—	—	57*
		1982	10	carbaryl(2)^A	12†
		1983	—	—	4
Lindenhurst	Lake	1981	—	—	154*
		1982	50	carbaryl(2)^A	2†
		1983	—	—	0
Morton	Tazewell	1981	—	—	3*
		1982	5	carbaryl(2)^G	0
		1983	—	—	0
Downers Grove	Du Page	1981	—	—	73
		1982	800	B.t.(2)^A	—
		1982	800	mass trapping	28
		1983	50	B.t.(2)^A	—
		1983	367	mass trapping	1
		1984	—	—	1
St. Charles	Kane	1981	—	—	5
		1982	—	—	26
		1983	90	B.t.(2)^A	1
		1984	40	mass trapping	1
Geneva	Kane	1985	25	mass trapping	?
Peoria	Peoria	1985	70	B.t.(3)	—
		1985	70	mass trapping	?
Aurora	Kane	1982	—	—	10
		1983	—	—	32
		1984	<7	B.t.(2)^G	0
Mundelein	Lake	1981	—	—	24
		1982	—	—	10
		1983	—	—	36
		1984	<7	B.t.(2)^G	1

Sources: N. C. Seaborg and E. S. Smith, Current status of the gypsy moth in Illinois. Div. of Agric. Industry, 1982; K. C. Kruse, Eradication of outlying gypsy moth infestations using *Bacillus thuringiensis* and mass trapping, USDA-APHIS-PPQ memorandum, 1984; Illinois Dept. Agric., Illinois gypsy moth summary, attachments to letter from S. E. Smith, plant and pesticide specialist supervisor, Ill. Dept. Agric., to R. Belhamm, Wash. Dept. Agric., Jan. 21, 1983; Illinois Dept. Agric., summary Illinois gypsy moth, 1983, memorandum; H. L. Ford, deputy director, PPQ-USDA, letter to W. Morris, director, Plant Health Div., Agriculture Canada, 1985; N. C. Seaborg, Div. of Agric. Industry Regulation, State of Illinois, phone conversation with S. H. Dreistadt, Dec. 12, 1984.

Notes: * =Other life stages found; † = Moths outside treatment area; (Number of applications); ^A = aerial application; ^G = ground application.

unteers distributed at three per acre. Eighty-four moths (concentrated around the properties of those who denied access to the B.t. sprayers) were captured in 1980. This was a 50 percent decrease from the previous year, when only one trap per acre was used. Two aerial applications of B.t. were made by the USDA in 1981, and no moths have since been found near the McHenry treatment area (USDA 1985).

A nearly identical B.t. and mass trapping program was implemented in Downer's Grove, Illinois, beginning in 1982. This infestation is now considered eradicated (Appelt 1985).

After the 1983 gypsy moth spray season, the agricultural advisor to the Illinois EPA wrote, "due to much public concern *Bacillus thuringiensis* is the only product currently used in [Illinois] government supervised sprayings" (A. G. Taylor, letter to R. Hirsch, 1983).

Wisconsin, 1975–1985. Wisconsin began limited (one hundred traps statewide) gypsy moth detection efforts in 1970. The first identified infestation was in the northeastern part of the state in 1975, when five moths were trapped in a rural area outside of Appleton. Subsequent moth captures (table 15.3) prompted a high density ("mass") trapping program during 1979, 1980, and 1981 that employed three traps per acre centered around an uninhabited ravine. No moths have been found within several miles of the treatment area since 1980. "Therefore, trapping, in concert with natural mortality factors is believed to have resulted in the elimination of the infestation" (Schwalbe 1981). According to the Wisconsin Department of Agriculture (1982), "This infestation is unique because it has apparently been eradicated using high density trapping only." Mass trapping alone has more recently been used with apparent success in Monona, Wisconsin, where no moths were detected during 1985 (K. P. Robert, letter to S. H. Dreistadt, 1985).

Wisconsin has applied selective controls over extended periods. All projects have employed mass trapping, usually in concert with applications of B.t., disparlure, Gypchek, or, to a very limited extent (eleven acres), carbaryl. According to the Wisconsin Department of Agriculture (1982), "The purpose of (these) pesticide treatments is to lower gypsy moth populations to below detectable levels. . . . Eradication, as such, is almost impossible to measure and is a philosophical term rather than an actuality."

Indiana, 1982–1985. The first gypsy moth infestations in Indiana were not identified until three decades after the discovery of extensive infestation in Michigan one hundred miles to the north and nearly a decade after infestations in neighboring Ohio and Illinois, in which an average of three times more eradication projects than in Indiana have been conducted. Even Wisconsin, located further from the general infestation, has treated more

Table 15.3   Gypsy Moth Eradication in Wisconsin (1975–1985)

City	County	Treatment Year	Acres	Method	Moths Caught
Appleton	Outagamie	1975	—	—	5
		1976	—	—	9
		1977	—	—	26
		1978	—	—	21
		1979	640	mass trapping	7
		1980	320	mass trapping	1†
		1981	160	mass trapping	0
		1982	—	—	3†
		1983–85	—	—	0
Oconomowoc Lake/ Okauchee	Waukesha	1977	—	—	5
		1978	—	—	643*
		1979	337	Gypchek(2)A	4
		1979	420	disparlureA	—
		1980	—	—	28*
		1981	10	carbaryl(1)G	—
		1981	200	mass trapping	74*
		1981	100	disparlure	—
		1982	300	mass trapping	19
		1983	385	mass trapping	21
		1984	35	B.t.(2)A	—
		1984	225	mass trapping	28*
		1985	2	B.t.(2)G	—
		1985	60	mass trapping	0
Delavan	Walworth	1981	—	—	14*
		1982	—	—	4
		1983	100	mass trapping	0
		1984–85	—	—	0
Elm Grove	Waukesha	1981	—	—	2
		1982	—	—	71*
		1983	60	B.t.(2)A	—
		1983	142	mass trapping	5
		1984	310	mass trapping	1†
		1985	—	—	0
Milwaukee	Milwaukee	1981	—	—	5
		1982	—	—	15
		1983	1	carbarylG	0
Monona	Dane	1981	—	—	236*
		1982	520	m.t./burlap bands	119*
		1983	220	m.t./burlap bands	44*
		1984	197	m.t./burlap bands	7
		1985	50	mass trapping	0

(continued)

Table 15.3   (*Continued*)

City	County	Year	Acres	Method	Moths Caught
				Treatment	**Moths Caught**
Hubertus	Washington	1983	—	—	39
		1984	160	mass trapping	10*
		1985	5	B.t.(2)^G	—
		1985	45	mass trapping	0
Sheboygan	Sheboygan	1983	—	—	18
		1984	—	—	8
		1985	39	mass trapping	0

Sources: Wis. Dept. of Agric., Proposed 1984 control of gypsy moth in Oconomowoc Township, Wisconsin, Final environmental impact assessment, 1984; K. P. Robert, administrator, Agric. Res. Management Div., Wis. Dept. Agric., letters to S. H. Dreistadt, Sept. 28, 1984, Nov. 18, 1985; USDA 1985; H. L. Ford, deputy director, PPQ-USDA letter to W. Morris, director, Plant Health Div., Agriculture Canada, 1985; Wis. Dept. Agric., Summing up gypsy moth control treatments in Wisconsin, 1975–1985, undated.

Notes: * = Other life stages found; † = Moth outside treatment area; (Number of applications); A = aerial application; G = ground application; disparlure = appalication of Hercon flakes; mass trapping (m.t.) = 3–9 traps/acre.

sites. It would be surprising if Indiana actually harbors so relatively few low-level infestations, given the number of identified infestations in adjoining states and the proximity of Indiana to long-established infestations, particularly in Michigan.

The first gypsy moth eradication project in Indiana employed two aerial applications of carbaryl over 300 acres in Terre Haute (Vigo County) in 1982. This was considered successful based on three seasons of negative gypsy moth trapping within the treatment zone.

Carbaryl was applied twice aerially in 1983 in Goshen (Elkhart County), Indiana, where 372 moths had been trapped the previous year. The post-treatment status of that infestation was characterized by the USDA as "eradicated," even though, "only 80 acres of 300 acres [were] treated. Moths [were] trapped on periphery of area" (USDA 1985). The area was mass trapped in 1984 (250 acres, 13 moths caught) and 1985 (500 acres, 8 moths caught). The state considers that this "infestation was not eliminated but greatly reduced" (J. J. Favinger, letter to S. H. Dreistadt, 1984), and no moths were detected during mass trapping in 1986 (Indiana Department of Natural Resources, county checklist of gypsy moth positive finds, undated).

During 1984 and 1985, infestations at five other sites in four counties (Allen, Bartholomew, Marion, and St. Joseph) were mass trapped (1045 acres). At two of these sites 455 acres were treated with *B.t.* (H. L. Ford, letter to W. Morris, 1985; D. Henry, letter to gypsy moth Science Advisory Panel, 1984).

Minnesota, 1983–1985.   Minnesota first records detecting egg masses and pupal cases on imported mobile homes, trailers, and nursery stock beginning in 1969. But the first breeding populations of gypsy moth were not found until more than fifteen years later in St. Paul and Woodbury. Although a lawsuit and lobbying by some local residents favored B.t., two aerial applications of carbaryl were made over 430 acres in 1983. Both infestations have been reported as eradicated.

The pathogen B.t. was sprayed twice over 82 acres at three sites in the St. Paul/Minneapolis area in 1984. No moths were captured after mass trapping on 146 acres at these locations (USDA 1985). An additional five sites in the St. Paul/Minneapolis area totaling 220 acres were sprayed with B.t. and were the sites of mass trapping in 1985 (ibid), and no further moths have been found there (D. Robertson, personal communication with S. H. Dreistadt, 1987).

Evaluation of Eradication Attempts in Great Lakes States.   Gypsy moth eradication is no different from most projects conducted in a "crisis" environment, in that records are often incomplete, inaccessible, or impenetrable. For projects in at least four of six Great Lakes states, state and federal records differ in reporting on the extent of the areas treated and whether the gypsy moth was eradicated. This is at least in part due to debates over how to define the zone of infestation when one or a few moths are captured some distance from the core area in which egg masses have been located and how to interpret moths found just inside or outside the boundaries of a treatment.

The best of these records have been summarized here and indicate that, through 1985, gypsy moth eradication projects were conducted during one or more years at 74 locations in Ohio (30), Illinois (19), Indiana (7), Minnesota (10), and Wisconsin (8). Ohio relied on aerial applications of carbaryl, whereas Wisconsin has employed a mix of selective tactics, primarily *Bacillus thuringiensis* combined with mass trapping. Illinois, Indiana, and Minnesota all switched from carbaryl to the use of *Bacillus thuringiensis* and mass trapping.

Each of these strategies appears to have been successful in most instances in eradicating local gypsy moth populations. It is difficult to assess, however, whether these ongoing efforts are successful statewide gypsy moth eradication programs. Each of these states has 4 million or more forested acres yet employs only several thousand detection traps. Trapping is concentrated where infestations are known or suspected, while outlying areas are much more lightly surveyed. Incomparable trapping methods are often used at sites treated with B.t. and carbaryl; fewer traps are typically used when carbaryl is applied because agricultural officials have generally believed that B.t. is less effective and must be employed in combination with mass trapping.

Eradication as a successful statewide policy requires the ability to readily detect low-level infestations, adequately define their extent, eliminate all gypsy moths from areas of defined infestations, and enforce quarantines sufficient to drastically reduce the rate at which gypsy moth is reintroduced and new infestations created. No Great Lakes state projects meet these requisites.

Most notable among these deficiencies is the inability to stem the influx of new gypsy moths. Introductions from outside represent a much greater source of new infestations than spread from the existing low-level infestations that are the focus of eradication activities. Effective quarantines would be prohibitive in cost, given the lack of natural barriers and the high volume of travel between Great Lakes states and the general infestation. The next major outbreak in the northeast will dramatically increase this already high rate of gypsy moth introduction.

In the absence of effective quarantines, the policy of "buying time" through the eradication of innocuous infestations is biologically difficult to justify as an effective method of preventing the gypsy moth from becoming established. Political and institutional support is insufficient to ensure the substantial and sustained resources that might allow a statewide policy of eradication to be technically feasible. Individual and organized citizens increasingly question the eradication approach and often oppose projects through lawsuits when chemical insecticides are proposed. Eradication in name only, most Great Lakes state programs more closely resemble suppression.

The gypsy moth management resources of the Great Lakes states should be allocated toward developing long-term strategies to monitor and selectively suppress populations that threaten to cause defoliation. A century of experience indicates that suppression or foliage protection of valuable trees, not eradication, is the only feasible long-term strategy for Great Lakes states attempting to minimize the adverse impacts of living with the gypsy moth.

Literature Cited

Appelt, P. 1985. A new eradication strategy for small, remote gypsy moth infestations. *J. Arboriculture* 11:242–46.

Beroza, M., E. C. Paszek, E. R. Mitchell, B. A. Bierl, J. R. McLaughlin, and D. L.

Chambers. 1974. Tests of a 3-layer laminated plastic bait dispenser for controlled emission of attractants from insect traps. *Environ. Entomol.* 3:926–28.

Brown, A. W. A. 1978. *Ecology of pesticides.* New York: Wiley and Sons.

Brown, A. W. A., and R. Pal. 1971. Insecticide resistance in arthropods. *World Health Org. Monogr.,* ser. 38, 2d ed.

Brown, J., D. Halliwell, and W. Gould. 1979. Gypsy moth defoliation impact in Rhode Island forests. *J. Forest.* 77:30–32.

Brown, W. L. 1961. Mass insect control programs: four case histories. *Psyche J. Entomol.* 68:75–111.

California Department of Food and Agriculture (CDFA). 1977. *Annual report, division of plant industry,* 23–27.

———. 1978. *Annual report, division of plant industry,* 47–49.

———. 1984. *Environmental Assessment of Gypsy Moth and Its Eradication in California, 1985 Program.* Revised November 1.

Campbell, R. W. 1976. Comparative analysis of numerically stable and violently fluctuating gypsy moth populations. *Environ. Entomol.* 5:1218–24.

———. 1979. Gypsy moth: Forest influence. *Agric. Inform. Bull.* 423. Washington, D.C.: USDA-Forest Service.

Campbell, R. W., D. Hubbard, and R. Sloan. 1975a. Patterns of gypsy moth occurrence within a sparse and numerically stable population. *Environ. Entomol.* 4:535–42.

———. 1975b. Location of gypsy moth pupae and subsequent pupal survival in sparse, stable populations. *Environ. Entomol.* 4:597–600.

Campbell, R. W., and R. Sloan. 1976. The influence of behavioral evolution on gypsy moth pupal survival in sparse populations. *Environ. Entomol.* 5:1211–17.

———. 1978. Numerical bimodality among North American gypsy moth populations. *Environ. Entomol.* 7:641–46.

Carson, R. 1962. *Silent spring.* Boston: Houghton Mifflin.

Commonwealth Institute of Entomology. 1981. *Distribution maps of pests.* Map. 26.

Cottam, C., and E. Higgins. 1946. DDT and its effect on fish and wildlife. *J. Econ. Entomol.* 39:44–52.

Daly, H., J. Doyen, and P. Ehrlich. 1978. *Introduction to insect biology and diversity.* New York: McGraw-Hill.

DeBach, P., ed. 1964. *Biological control of insect pests and weeds.* New York: Reinhold.

Dreistadt, S. H. 1983a. An assessment of gypsy moth eradication attempts in Michigan (Lepidoptera: Lymantriidae). *Great Lakes Entomol.* 16:143–48.

———. 1983b. Gypsy moth in California: Is carbaryl the answer? CBE *Environ. Rev.,* May/June, 3–7.

Dreves, J. 1982. Gypsy moth in Michigan: Successes and problems. In *Urban and suburban trees: Pest problems, needs, prospects and solutions,* proceedings of a conference held at Michigan State University, April 18–20. East Lansing: Kellogg Center for Continuing Education.

Dubois, N. R. 1981. Microbials. In *The gypsy moth: Research toward integrated pest management,* ed. C. C. Doane and M. L. McManus, 444–53. *U.S. Dept. Agric. Tech. Bull.* 1584.

Dunlap, T. 1980. The gypsy moth: A study in science and public policy. *J. Forest History* 24:116–26.

————. 1981. DDT: *Scientists, citizens, and public policy*. Princeton: Princeton University Press.

Elkinton, J. S., and R. T. Cardé. 1981. The use of pheromone traps to monitor distribution and population trends of the gypsy moth. In *Management of insect pests with semiochemicals*, ed. E. R. Mitchell, 41–55. New York: Plenum Press.

Etkind, P. H., T. M. ODell, A. T. Canada, S. K. Shama, A. M. Finn, and R. Tuthill. 1982. The gypsy moth caterpillar: A significant new occupational and public health problem. *J. Occup. Med.* 24:659–62.

Felt, E. P. 1942. The gypsy moth threat in the United States. *Eastern Plant Board Circ.* 1:1–16.

Forbush, E. H., and C. H. Fernald. 1896. The gypsy moth, *Porthetria dispar* (Linn.). Boston: Wright and Potter.

Gerardi, M. H., and J. K. Grimm. 1979. The history, biology, damage, and control of the gypsy moth *Porthetria dispar* (L.). London: Associated University Press.

Giese, R., and M. Schneider. 1979. Cartographic comparisons of Eurasian gypsy moth distribution (*Lymantria dispar*, Lepidoptera: Lymantriidae). *Entomol. News* 90:1–16.

Ginsburg, R. 1981. Let us (S)pray? CBE *Environ. Rev.*, July/August, 3–6.

Glaser, R. W. 1915. Wilt of gypsy moth caterpillars. *J. Agr. Res.* 4:101–28.

Hanna, M. 1982. Gypsy moth (Lepidoptera: Lymantriidae): History of eradication efforts in Michigan, 1954–1981. *Great Lakes Entomol.* 15:191–98.

Henry, D. 1981. Emergency and Special Projects, California Department of Food and Agriculture (CDFA). Memorandum to I. Siddiqui. September 28.

Hough, J., and D. Pimentel. 1978. Influence of host foliage on development, survival and fecundity of the gypsy moth. *Environ. Entomol.* 7:97–102.

Houston, D. R. 1979. Classifying forest susceptibility to gypsy moth defoliation. *U.S. Dept. Agric. Handbk* 542.

Hoy, M. 1976. Establishment of gypsy moth parasitoids in North America: An evaluation of possible reasons for establishment or non-establishment. In *Perspectives in forest entomology*, ed. J. F. Anderson and H. K. Kaya, 215–32. New York: Academic Press.

Jacobson, M., M. Schwarz, and R. M. Waters. 1970. Gypsy moth sex attractants: A reinvestigation. *J. Econ. Entomol.* 63:943–45.

Jobin, L. 1982. Results of aerial treatments with Dimilin and *Bacillus thuringiensis* for gypsy moth (*Lymantria dispar* [L.]) control in Quebec. *Can. For. Serv. Res. Notes* 2:18–20.

Knipling, E. F. 1979. The basic principles of insect population suppression and management. *U.S. Dept. Agric. Handbk* 512.

Lewis, F. B., and W. G. Yendol. 1981. Efficacy. In *The gypsy moth: Research toward integrated pest management*, ed. C. C. Doane and M. L. McManus, *U.S. Dept. Agric. Tech. Bull.* 1584.

Maryland State Department of Agriculture (MSDA). 1983. Work plan: IPM pilot project for gypsy moth in Maryland.

Massachusetts Board of Agriculture (MBA). 1890 through 1903, each year. Annual Reports on gypsy moth work.

Massachusetts Department of Environmental Management (MDEM). 1982. Gypsy moth policy–1982. Division of Forests and Parks.

Mastro, V. C. 1984. Sterile male trial, Berrien County, Michigan. In *Otis methods development center report,* USDA-APHIS, 37–38.

Mastro, V. C., and T. M. ODell. 1985. Induced inherited sterility trial, Horry County, South Carolina. In *Otis methods development center report,* USDA-APHIS, 83.

Mastro, V. C., C. P. Schwalbe, J. D. Tang, and D. R. Lance. 1985. Sterile F_1 demonstration project, Darke County, Ohio. In *Otis methods development center report,* USDA-APHIS, 108–111.

McManus, M. L., and T. McIntyre. 1981. Introduction. In *The gypsy moth: Research toward integrated pest management,* ed. C. C. Doane and M. L. McManus, 1-7. *U.S. Dept. Agric. Tech. Bull.* 1584.

Michigan Department of Agriculture (MDA). 1976. *Gypsy moth treatment program 1976.* Environmental impact statement supplement no. 1.

————. 1984. *State of Michigan 1984 gypsy moth management program.* Environmental statement and site specific environmental analysis. March.

Miller, J. C., and K. J. West. 1987. Efficacy of *Bacillus thuringiensis* and diflubenzuron on Douglas-fir and oak for gypsy moth control in Oregon. *J. Arboriculture* 13:240–42.

Moore, R. 1983. Gypsy moth eradication treatments in Michigan. Memorandum from USDA-APHIS district director to Ken Kruse, area director, March 30.

Moore, T. E., and M. Hanna. 1984. Gypsy moth distribution in Michigan. *Jack-Pine Warbler* 62:1–5.

Morse, J., and G. Simmons. 1978. Alternatives to the gypsy moth eradication program in Michigan. *Great Lakes Entomol.* 11:243–48.

National Research Council. 1975. *Pest control: An assessment of present and alternative technologies.* Volume 4. Washington, D.C. National Academy of Sciences.

Neher, L., R. Segawa, and R. Oshima. 1982. *Monitoring of the 1982 gypsy moth eradication ground spray program in Santa Barbara County.* CDFA.

New Jersey Department of Agriculture, (NJDA). 1932, 1935. *Seventeenth annual report,* 118–21; *Twentieth annual report,* 148–52.

New York Department of Environmental Conservation (NYDEC). 1983. Cooperative gypsy moth integrated pest management project. Appendix B.

Nichols, J. O. 1961. The gypsy moth in Pennsylvania: Its history and eradication. *Penn. Dept. Agric. Misc. Bull.* 4404.

————. 1980. *The gypsy moth.* Commonwealth of Pennsylvania Department of Environmental Resources.

Pennsylvania Department of Environmental Resources (PDER). 1982. Fact sheet on the microbial insecticide, *Bacillus thuringiensis (B.t.),* and its use in gypsy moth suppression.

Perry, C. C. 1955. *Gypsy moth appraisal program and 1952 proposed plan to prevent spread of moths, U.S. Dept. Agric. Tech. Bull.* 1124.

Pimentel, D. 1971. *Ecological effects of pesticides on non-target species.* Washington, D.C. Executive Office of the President, Office of Science and Technology.

Popham, W. L., and D. G. Hall. 1958. Insect eradication programs. *Ann. Rev. Entomol.* 3:335–54.

Rachel Carson Council. 1987. Coping with the gypsy moth: A list of pesticides registered for use in control of the gypsy moth provided by the U.S. Environmental Protection Agency.

Raimo, B. J., and R. C. Reardon. 1981. Preliminary attempts to use parasites in combination with pathogens in an integrated control approach. In *The gypsy moth: Research toward integrated pest management,* ed. C. C. Doane and M. L. McManus 408–12. *U.S. Dept. Agric. Tech. Bull.* 1584.

Reardon, R. C. 1981. Parasite/gypsy moth interactions: Summary and suggested areas of future research. In *The gypsy moth: Research toward integrated pest management,* ed. C. C. Doane and M. L. McManus, *U.S. Dept. Agric. Tech. Bull.* 1584.

Respicio, N. P., and A. J. Forgash. 1977. Insecticide susceptibility in New Jersey gypsy moth (Lepidoptera: Lymantriidae) populations. *J. N.Y. Entomol. Soc.* 85:56–60.

Schwalbe, C. 1981. Evaluation of mass trapping for control, Appleton, Wisconsin. In *Otis methods development center report,* USDA-APHIS.

Shama, S. K., P. H. Etkind, T. M. ODell, A. T. Canada, A. M. Finn, and N. A. Soter. 1982. Gypsy-moth-caterpillar dermatitis. *N. Engl. J. Med.* 306:1300–301.

Sheals, R. 1946. New developments in gypsy moth control. *Agric. Chem.* 1:22–25.

Siddiqui, I. A. 1984. Pest Detection/Emergency Projects, CDFA Memorandum to Gypsy Moth Science Advisory Panel. September 10.

Smilovitz, Z., and L. D. Rhodes. 1972. Parasites of gypsy moth located in Pennsylvania. Penn. State Univ. College Agric. Expt. Sta., *Science in Agriculture* 19:4.

Smith, O. 1976. Changes in eastern forests since 1600 and possible effects. In *Perspectives in forest entomology,* ed. J. F. Anderson and H. K. Kaya, 3–20. New York: Academic Press.

Stephens, G. 1976. The Connecticut forest. In *Perspectives in forest entomology,* ed. J. F. Anderson and H. K. Kaya, 21–30. New York: Academic Press.

———. 1981. Defoliation and mortality in Connecticut forests. *Conn. Agric. Expt. Sta. Bull.* 796.

Sumnick, W. E. 1958. Can we survive the gypsy moth spray? *Amer. Bee J.:* (June) 224–25.

M. Teta. Undated. The McHenry Story. McHenry County Defenders, Crystal Lake, Illinois.

Tomlin, A. D., and A. J. Forgash. 1972. Toxicity of insecticides to gypsy moth larvae. *J. Econ. Entomol.* 65:953–54.

Troast, R., and J. Sanders. 1979. *Diflubenzuron decision document.* Washington, D.C.: U.S. Environmental Protection Agency Special Pesticide Review Division. 26 March.

U. S. Department of Agriculture (USDA). 1982. Eradication experiences with isolated gypsy moth infestations—1967–1981. APHIS-PPQ.

———. 1985. *Gypsy moth suppression and eradication projects.* Final environmental impact statement. Washington, D.C.

U. S. Department of Commerce (USDC). 1975. Historical statistics of the United States, 125. Washington, D.C.

———. 1982. Statistical abstract of the United States 103rd ed., 451. Washington, D.C.

U. S. Environmental Protection Agency (USEPA). 1980. Carbaryl decision document.

Waller, W. K. 1958. Poison on the land. *Audubon,* March–April, 68–70.

Wallner, W. E., and W. L. McManus. 1981. Summary [of population dynamics]. In *The gypsy moth: Research toward integrated pest management*, ed. C. C. Doane and M. L. McManus, 202–203. *U.S. Dept. Agric. Tech. Bull.* 1584.

Weis, P., and J. S. Weis. 1979. Congenital abnormalities in estuarine fishes produced by environmental contaminants. pp. 94–106. In *Animals as monitors of environmental pollutants,* Symposium on Pathobiology of Environmental Pollutants: Animal Models and Wildlife as Monitors. Univ. Conn., 1977. Washington D.C.: National Academy of Sciences.

Weiss, H. B. 1921 through 1931, each year. Results of each fiscal year's work against the gypsy moth in New Jersey. *N.J. Dept. Agric. Circ.* 38, 56, 67, 79, 89, 105, 127, 150, 169, 189, 208.

White, W. B., W. H. McLane, and N. F. Schneeberger. 1981. Pesticides. In *The gypsy moth: Research toward integrated pest management*, ed. C. C. Doane and M. L. McManus, 423–42. *U.S. Dept. Agric. Tech. Bull.* 1584.

White, W. B., and N. Schneeberger. 1981. Socioeconomic impacts. In *The gypsy moth: Research toward integrated pest management*, ed. C. C. Doane and M. L. McManus, 681–94. *U.S. Dept. Agric. Tech. Bull.* 1584.

Wisconsin Department of Agriculture (WDA). 1982. *Environmental impact assessment for the proposed 1982 control of gypsy moth in Monona, Wisconsin,* A-2.

Wolfe, R. D. 1985. Gypsy moth, a regional nuisance becomes a national dilemma. In *Insect and disease conditions in the United States, 1979–83,* ed. R. C. Loomis, S. Tucker, and T. H. Hofacker, 8–15. USDA-Forest Service. General Technical Report WO–46.

Worrell, A. C. 1960. Pests, pesticides, and people. *Amer. Forests* 66:39–81.

16

Yellow Fever Mosquito

Durward D. Skiles

The history of the war of eradication waged against the mosquito *Aedes aegypti* (L.) is inseparable from the history of man's attempt to control yellow fever, a viral disease for which *Ae. aegypti* is the primary vector within human populations. *Ae. aegypti* is, however, also the primary vector of the viral disease dengue and is a demonstrated or suspected vector of several other diseases of man and animals. Because the last documented case of yellow fever transmitted by *Ae. aegypti* in the Western Hemisphere occurred in Trinidad in 1954 (Schliessmann and Calheiros 1974), recent outbreaks of dengue have been the major stimulus of renewed efforts to eradicate *Ae. aegypti*. Nevertheless, the threat of epidemic "urban yellow fever" in the human population remains very real, owing to the continued presence of *Ae. aegypti* throughout the tropical and subtropical regions of the world (fig. 16.1) and the existence of a cycle of "jungle yellow fever" among various populations of new world monkeys and probably other mammals as well.

Following the identification of *Ae. aegypti* as an efficient vector of yellow fever by Reed and his coworkers in Cuba in 1901, vigorous campaigns to free heavily infested areas of the mosquito were soon begun in several countries. The spectacular successes of Gorgas in Cuba and Panama are well known. Those successes clearly demonstrated that proper control of *Ae. aegypti* was tantamount to eradication of yellow fever. As a result, in 1915 the International Health Commission of The Rockefeller Foundation resolved to eradicate the disease from its endemic areas and, thereby, from the world (Warren 1951).

In fewer than twenty years, however, the goal of world-wide yellow fever eradication was abandoned with the discovery of a sylvatic monkey-mosquito-monkey cycle of yellow fever that existed entirely independent of both man and *Ae. aegypti*. Thus, a single person could contract yellow fever in the jungle and return to an urban area where *Ae. aegypti* could spread the disease among the human population. That discovery led Soper and Wilson (1942) to propose that the goal of disease eradication via maintenance of *Ae. aegypti* populations at very low levels be replaced by the goal of

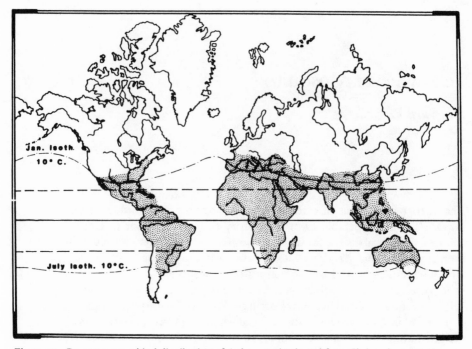

Figure 16.1. Present geographical distribution of *Aedes aegypti* (adapted from Christophers 1960).

regional species eradication. They reasoned that over the long term an intense eradication program of finite duration was less costly than a program of suppression of indefinite duration.

In 1947, to protect countries attempting to eradicate *Ae. aegypti* from reinfestation, the Pan American Health Organization (PAHO) resolved to eradicate *Ae. aegypti* from the hemisphere. In fewer than ten years, Severo (1956) was able to report that eradication of *Ae. aegypti* was either accomplished or imminent in fourteen countries in Central and South America. In the same year, the Operational Manual of PAHO predicted that with the aid of DDT anti-*aegypti* campaigns were assured of success. Between 1958 and 1965 eradication was confirmed in seventeen of the twenty-three initially infested countries, which comprised 73.5 percent of the initially infested land area of the hemisphere.

But eradication was accomplished in only two of twenty-six Caribbean countries and territories, and since 1965 reinfestations have occurred in at least seven of the seventeen countries that had achieved eradication (Schliessmann and Calheiros 1974). Cuba and the United States, two countries engaging in considerable commerce with the rest of the hemisphere, have never eradicated *Ae. aegypti*. In Cuba the population of *Ae. aegypti* was sufficiently large to produce epidemics of dengue in 1977 and 1981 that comprised 900,000 reported cases and perhaps ten times as many unreported cases (Tonn, Figueredo, and Uribe 1982).

While the 1977 epidemic was apparently allowed to run its course, the 1981 epidemic was quickly halted with an intense, country-wide anti-*aegypti* program involving breeding site elimination and pesticide application. But this recent success, like the earlier successes of Gorgas and Soper, is only a single battle in a protracted war with an unclear outcome. *Ae. aegypti* has demonstrated the capacity to develop resistance to pesticides, including DDT, malathion, and fenthion (ibid; Fox 1973, 1977). Whereas *Ae. aegypti* in the Western Hemisphere was previously considered to breed almost exclusively in man-made containers in urban areas, there are now increasing numbers of reports of breeding in rural areas and in such natural sites as tree holes and rock depressions. Furthermore, the tendency of an increasingly affluent and mechanized human society to dispose of its rubbish (especially bottles, cans, and auto tires) above ground has created a rich array of *Ae. aegypti* breeding sites that were unavailable to the mosquitoes of the times of Gorgas and Soper. Despite surveillance programs, man remains a very effective dispersal agent for *Ae. aegypti,* primarily by boats ranging from large commercial vessels to small unchartered craft. Such conditions have led Tonn, Figueredo, and Uribe (1982), in their review of the current status of *Ae. aegypti,* yellow fever, and dengue in the Americas, to remark "Today, *Ae. aegypti* is still winning over the efforts of man despite new control equipment and insecticides as well as millions of dollars being spent annually against it."

Life History and Habits

The importance of *Ae. aegypti* as a vector of yellow fever in human populations is due directly to the close association of the species with humans. Particularly in the Western Hemisphere it breeds primarily in water that has collected in man-made containers of almost any kind, both in houses and out-of-doors. These include containers actively employed for water collection or storage (cisterns, flower pots, water jugs, antiformicas, animal watering troughs), discarded objects (pots, cans, bottles, auto tires, auto bodies, drums), masonry defects, and neglected household items (gutters, drain traps, unused toilets).

Because adults usually stay within a few hundred yards of their larval site, long range dispersal of the species depends primarily upon man. As a result, *Ae. aegypti* moves along human travel and trade routes and is particularly prevalent along coastal areas and harbors. The abundance of bilge water and the necessity of transporting fresh water on ships ensures species dispersal along shipping routes. Tonn Figueredo, and Uribe (1982) reported that almost every month at least one ship entering Panamanian ports if found to be infested with *Ae. aegypti*. Hence eradication alone is not sufficient to ensure freedom from infestation in a given locality. Continual surveillance is required to guard against reintroduction.

It is generally assumed that *Ae. aegypti* was originally a forest dwelling species breeding in tree holes and other natural water containers, as it often does in Africa today. Until recently, however, the form found in the Western Hemisphere was thought to breed almost exclusively in association with humans (Christophers 1960; Taylor 1951), and any occurrence of immatures in natural water containers was considered aberrant and unlikely to perpetuate the local population. But, as Tonn, Figueredo, and Uribe (1982) noted with concern, "More and more countries are reporting the occasional breeding in natural containers such as tree and rock holes." Schliessmann (1967) reported that even in the United States larvae were occasionally found in tree holes and leaf axils of bromeliades and travelers' palms.

The existence of *Ae. aegypti* populations remote from human population centers has obvious and important implications for eradication efforts. First, confinement of eradication operations to regions inhabited or frequented by humans would not ensure eradication of the species. Second, eradication efforts of the classical kinds—breeding-site elimination and regional pesticide application—are not effective against populations in large tracts of dense forest or jungle. A successful effort would probably require some biological control agent that could search out and destroy populations of *Ae. aegypti*. The threat posed by nonurban reservoirs of *Ae. aegypti* and yellow fever is illustrated by the 1971 outbreak of yellow fever in Luanda, Angola (Ribeiro 1973). Since the 1860–1872 epidemic, there had been no clinical cases of yellow fever, even though surveys of immunity showed that the virus continued to be active in the northern part of Angola. In 1970 *Ae. aegypti* invaded Luanda via breeding sites provided by an enormous number of discarded auto tires around the city, and the epidemic ensued.

The life cycle of *Ae. aegypti* comprises the egg, four larval instars, the pupa, and the adult. Successful reproduction requires male fertilization of the female, a blood meal for the female, suitable water for breeding, and temperatures that will permit activity and maturation of egg and larva. The life cycle requires about sixteen to thirty days in the tropics and perhaps forty to forty-five days in temperate regions (Christophers 1960). Under favorable conditions the life cycle may be completed in ten days. The tropical life cycle is roughly as follows: oviposition to egg hatch, three days; egg hatch to adult, ten days; pupal eclosion to first blood meal, three days; and egg maturation after blood meal, three days.

Breeding occurs whenever conditions are favorable, usually when temperatures are between 68° and 102°F. Adults become inactive below about 60°F and are rapidly killed below 32°F. The eggs can survive desiccation for months (420 days in the laboratory according to Severo [1956]), but cannot survive subfreezing temperatures for more than a few days. Therefore, *Ae. aegypti* can remain active and breed throughout the year only in the tropics; they become inactive in colder months in temperature regions where populations survive the winter because of protected locations.

Although adults have survived in the lab for more than one year, the natural survival time is probably only a few weeks. Taylor (1951) noted, however, that significant numbers of adults probably survive for six to eight weeks, since that is the usual time required to suppress a yellow fever epidemic after breeding of the mosquito has been curtailed by control measures. The female lays up to 50 eggs per blood meal. Most (70 percent) of the females probably die after the first oviposition but may survive to breed again. One female was observed to lay 752 eggs in seventy-two days in the laboratory (Christophers 1960). Hence, under favorable conditions, the importation of a few gravid females could lead to a population of millions of mosquitoes in a month or so.

After the yellow fever virus is injected into a human during a blood meal of *Ae. aegypti,* three to six days are required for incubation. The virus circulates in the blood of the human host for no more than four to five days, and it is only during that brief period that a noninfective mosquito can pick up the virus during a blood meal. It is eight to twelve days before this mosquito then becomes infective and can transmit yellow fever. Once infective, the adult mosquito remains so for the rest of her life.

Yellow fever can remain endemic only in regions where adult *Ae. aegypti* are active throughout the year. But the disease is readily transported to nonendemic regions by infected humans or mosquitoes. If an infective person enters a region containing both the vector and susceptible humans, it will be two weeks or more before the first secondary cases of yellow fever appear. But if infective mosquitoes are imported, the first secondary cases can appear in less than one week. Although control measures have eliminated yellow fever as a serious health problem in the Western Hemisphere, the threat of renewed epidemic disease activity is very real wherever large population densities of *Ae. aegypti* exist.

Control

The 1901 campaign against *Ae. aegypti* in Havana, under the direction of Major W. C. Gorgas of the United States Army Medical Corps, demonstrated that yellow fever could be eliminated from an urban area by suppressing the population of *Ae. aegypti*. In the last decade of the nineteenth century there was an average of more than five hundred deaths per year from yellow fever in Havana. The anti-*aegypti* campaign was begun in February 1901. The last case of yellow fever occurred in September 1901. In 1905 an outbreak of yellow fever occurred but was quickly suppressed. In 1909 the entire island of Cuba was declared free of yellow fever.

In 1903 a campaign against *Ae. aegypti* was begun in the city of Rio de Janeiro by Oswaldo Cruz. The elimination of yellow fever was completed in about seven years. The disease did not recur until the outbreak of 1928. In another anti-*aegypti* campaign, begun in November 1918, in Guayaquil, Ecuador, the disease was apparently eliminated within one year (Warren 1951).

The primary control method in these early campaigns was breeding-site elimination. Neglected collections of water were eliminated or covered with oil. Bottles and cans were cleaned up, cisterns were screened, and householders were required to maintain their premises free of mosquito larvae.

Predatory fish also played an important role in controlling *Ae. aegypti*. Households, particularly those in which yellow fever had recently occurred, were often fumigated with the insecticide of the time, (for example, sulfur dioxide, pyrethrum, or tobacco smoke [Smith 1951]). Government inspectors made regular tours through the community to ensure that control was maintained.

Today, breeding-site elimination remains at the core of any control program, along with surveillance to prevent reinfestations. But since the 1950s widespread applications of synthetic pesticides have formed an integral part of control programs. In 1956 the Operational Manual of PAHO declared that DDT would insure the success of anti-*aegypti* campaigns (Tonn, Figueredo, and Uribe 1982).

Although initially effective, the use of synthetic pesticides quickly led to the development of pesticide-resistant strains of *Ae. aegypti*. During the 1954 outbreak of yellow fever in Trinidad, an island-wide application of 75 percent wettable powder DDT at the rate of one drop per gallon of breeding site water failed to eradicate *Ae. aegypti* in even a single area. Later laboratory investigations showed that larvae could survive DDT concentrations of up to 200 parts per million for forty-eight hours, "thus shattering the position of [*Ae. aegypti*] as the classic susceptible to the chemical" (Gillette 1956).

Whereas Eliason, Kilpatrick, and Babbitt (1970) concluded that weekly and biweekly aerial ultra low volume (ULV) applications (3 fluid ounces per acre) of malathion were effective in the control of *Ae. aegypti* in Florida, Fox (1980) found that in Puerto Rico the aerial and ground spraying program of 1978 was ineffective. In fact, Fox found a considerable increase in larval and pupal numbers after the spray program! Although some of the failure was attributed to the probable failure of the malathion spray to reach its target, it appeared likely that malathion resistance also played a role, because Fox (1973) had already demonstrated that under continuous selection pressure on adult *Ae. aegypti* for twelve generations in the laboratory, malathion resistance increased twelve-fold. Furthermore, Fox and Bayona (1972) found wild malathion-resistant strains of *Ae. aegypti* after an intensive malathion campaign in Puerto Rico. "Following an epidemic of dengue in 1963, a large scale eradication program involved the spraying of hundreds of thousands of gallons of 2.5 percent solution (of malathion) over most of the Island at a cost of millions of dollars. The results were not impressive. In 1969, after five years of effort, *Ae. aegypti* was abundant everywhere, another epidemic of dengue broke out, and malathion continued to be applied in huge quantities." Fox (1977) also found increased resistance to fenthion, an effective larvicide, when *Ae. aegypti* larvae were subjected to selection pressure in the laboratory.

Christophers (1960) listed an array of natural enemies attacking all stages in the life cycle of *Ae. aegypti*. However, Christophers stated that larvivorous fish were "the only natural enemy that has been extensively used with success in the control of mosquito breeding (including *Ae. aegypti*) as a sanitary measure." Indeed, larvivorous fish were "probably the most effective of all natural enemies." For example, Smith (1951) noted that during the early 1920s in Mérida, Mexico, Yucatan fish were used to free cisterns of *Ae. aegypti* infestations. "In May 1921, Dr. M. E. Connor reported that in 12,324 water containers in which fish were used, inspection failed to reveal a single one harboring larvae or pupae." But although species of *Gambusia* and other fish have been used successfully in cisterns and wells, they cannot be used for permanent control in small, ephemeral collections of water or in areas where an alternate food source is not present.

The predaceous larvae of certain species of mosquitoes can also be effective in controlling *Ae. aegypti* larvae. For example, Christophers noted that a single larva of *Megarhinus* (also known as *Toxorhynchites*) can devour more than twenty *Ae. aegypti* larvae in a single night. In a field trial, Gerberg and Visser (1978) found that larvae of the predatory mosquito *Toxorhynchites brevipalpis* Theobald was effective in controlling *Ae. aegypti* and had the advantage that it could be easily reared locally. From laboratory trials, Ignoffo et al. (1980) found that *Ae. aegypti* was highly susceptible to four varieties of *Bacillus thuringiensis*. Nevertheless, recent reviews of *Ae. aegypti* eradication efforts (National Academy of Sciences 1976; Tonn, Figueredo, and Uribe 1982) contain no mention of biological control and reveal that, apart from breeding-site reduction, pesticide application is the primary control method.

History of Eradication Efforts

The history of the effort to eradicate yellow fever and *Ae. aegypti* provides a valuable perspective on present approaches to control and eradication.

Given that yellow fever was one of the most dreaded diseases in the world (Cloudsley-Thompson 1976) and became the most dreaded disease in the Americas during the eighteenth and nineteenth centuries (Smith 1951), it is hardly surprising that medical and health officials and a prominent philanthropic organization would be captivated by the prospect of eradicating the disease. Neither is it surprising that when the dream of disease eradication vanished it was quickly replaced by a dream of disease prevention through vector eradication. Notable local and regional successes and the invention of DDT and other potent synthetic insecticides made the attainment of those goals appear inevitable to many reasonable persons. The history of the *Ae. aegypti* is replete with examples of extreme reluctance to abandon established concepts, even though much new evidence invalidated those concepts.

The early campaigns established clearly that it was not necessary to completely eradicate the arthropod vector to eliminate yellow fever from a particular region. When *Ae. aegypti* populations were suppressed below a critical density, the disease was no longer transmitted and soon disappeared. Roughly speaking, when the premises breeding index (the percentage of houses harboring one or more water containers with *aegypti* larvae therein) was lowered to 5 percent or less, yellow fever was quickly eliminated. This was indeed fortunate, for as Smith (1951) noted it was relatively easy to lower the premises index to below 5 percent, but reduction of the index to below 1 or 2 percent was much more difficult and expensive.

Impressed by the early successes and secure in the belief that the epidemiology of yellow fever was relatively simple, the International Health Commission of The Rockefeller Foundation resolved in 1915 (Warren 1951) that it was "prepared to give aid in the eradication of this disease in those areas where the infection is endemic and where conditions would seem to invite cooperation for its control." From subsequent accounts written by those intimately involved in the eradication effort (Strode 1951), it is clear that for approximately the next two decades anti-*aegypti* efforts were motivated by a "dream of final eradication of yellow fever" (Soper et al. 1943), based on the beliefs that yellow fever could be contracted only via the bite of *Ae. aegypti*, that there were a strictly limited number of endemic foci (known as "key centers") of the disease that served as seedbeds for all cases of the disease throughout the world, and that eradication of the disease from those few foci would forever rid the world of yellow fever. Indeed, by 1925 the disease was considered to have been eliminated from Mexico, Central America, and Ecuador, and efforts in other endemic regions were meeting with success (Warren 1951).

The disease eradication effort received a rude shock in 1928 with the outbreak of yellow fever in Rio de Janeiro after a hiatus of twenty years. The outbreak was particularly perplexing, since the nearest known infected area in Brazil was almost one thousand miles away (Smith 1951). Shortly there-after, in 1932, a yellow fever outbreak occurred in the Valle do Chanaan, Espirito Santo, Brazil—a rural area outside the known endemic region of the disease. A thorough investigation failed to turn up any *Ae. aegypti* in the infected region (Soper et al. 1933). The implications of these outbreaks were not fully appreciated, however, since Soper et al. (1933) concluded, "This (Valle do Chanaan) epidemic was apparently self-limited, and it is possible that there are no rural regions in America that are truly endemic for yellow fever." Clearly, it was not easy to abandon, or even alter, a concept that had until then worked so well.

But during the next few years extensive immunity surveys revealed the "vast silent reservoir of infection" in the interior of South America (Soper 1937), Central America, and Africa (Smith 1951). Interestingly, immunity against yellow fever was found unexpectedly to be so widespread that the validity of the serological immunity test was called into question (ibid). Again, the entrenched epidemiological concept of yellow fever was not easily abandoned. In 1934 F. L. Soper conceded the possibility of vertebrate yellow fever hosts other than man (Taylor 1951) and soon thereafter used the term "jungle yellow fever" to apply to yellow fever contracted by man in the forest in the absence of *Ae. aegypti* (Soper 1935, 1936). By the late 1930s the concept of a forest monkey-mosquito-monkey cycle of yellow fever was generally accepted. It had been demonstrated that animals other than man could contract and die from yellow fever, that several species of *Aedes* and at least one species of *Haemagogus* could transmit the virus among monkeys, and that the bites of species of mosquitoes other than *Ae. aegypti* captured in the jungle could infect monkeys with yellow fever. With that knowledge came the abandonment of the dream of eradicating yellow fever.

The outbreak of yellow fever in Rio de Janeiro in 1928 had produced the realization that relaxation of *Ae. aegypti* control efforts could have disas-trous consequences, since *Ae. aegypti* populations increased rapidly in the absence of control. That knowledge, coupled with the discovery of endemic jungle yellow fever, led Soper and Wilson (1942) to declare that the solution was species eradication, which they defined as "the complete extermination of the species under consideration in all phases of its development: ovum, larva, pupa, and imago. When species eradication has been accomplished, the crucial test of discontinuation of all control measures, in the presence of suitable breeding places, is not followed, in the absence of reimportation, by reappearance of the species."

Soper et al. (1943) went further and provided an economic justification of species eradication. "Fortunately the difference in expense between maintaining a safe density of *Ae. aegypti* in a city and keeping the same city completely free of this species is so great that the cost of clearing contiguous areas of the mosquito to protect a zero index is more than justified over a period of years. *Thus any program of species eradication tends to expand automatically to include even the smallest and most distant places, until the entire region is cleared of the species.*" (emphasis added).

The latter statement identifies the Achilles heel of any attempt at permanent, large-scale eradication of *Ae. aegypti*. Nevertheless, Soper et al. (1943) proudly stated, "The year 1940 saw the return to government control at government expense of a streamlined National Yellow Fever Service operating throughout the entire country, conscious of its direct responsibility for the occurrence of any *Ae. aegypti*-transmitted yellow fever, and willing to declare its program to be the eradication of *Ae. aegypti* from Brazil."

The permanent Brazilian campaign consisted of three phases: (1) the initial clean-up and elimination of easily accessible breeding foci (this phase was essentially similar to the earlier emergency suppression campaigns that rid localities of yellow fever but not of *Ae. aegypti*); (2) the discovery and elimination of residual, hidden, or inaccessible breeding foci that were responsible for the continued presence of *Ae. aegypti* after phase one; (3) the permanent maintenance of a sentinel service to discover and eliminate such reinfestations as might occur.

The discovery and elimination of residual foci in phase two was accomplished by persistent efforts to find breeding foci in areas where *Ae. aegypti* continued to occur (the most effective method to pinpoint the focus was to capture adults and map the capture sites) and either to destroy or pour oil on those found.

The Brazilian anti-*aegypti* campaign was carried out in a military fashion. All premises were visited regularly by uniformed inspectors; inspectors were held strictly accountable for their work; detailed forms were filled out, and permanent records kept; householders were required by law to maintain their premises free of breeding mosquitoes. In 1946 DDT was introduced into the campaign and used initially as a residual spray on the walls of houses, a practice that was soon found to be expensive and unnecessary. Later, a DDT emulsion was sprayed on household water containers and other collections of water where breeding was prevalent. The campaign was highly successful. Smith (1951) was able to state, "By the end of 1949 *Ae. aegypti* had been eradicated from 16 states, five territories, and the Federal District of Brazil. A determined attempt is being made to eliminate the vector from the few remaining foci in the northeastern section of the country."

As noted previously, to protect countries trying to eradicate *Ae. aegypti* from reinfestation, the member nations of the PAHO resolved in 1947 to eradicate *Ae. aegypti* from the Western Hemisphere. In 1956 Severo reported that *Ae. aegypti* eradication was accomplished or imminent in fourteen countries in Central and South America. At the same time, he criticized the United States for being the only country that had not begun an eradication campaign. In 1963 Soper resumed his call for the eradication of *Ae. aegypti.* "The well-advanced program of *Ae. aegypti* eradication in the Americas has almost entirely eliminated the threat of urban yellow fever. For the complete and permanent elimination of this threat it is essential that those countries still harboring the *Ae. aegypti* mosquito proceed forthwith with its eradication. The principal delinquent in this international effort is the United States of America." (The United States joined the eradication effort in 1964, using DDT as the insecticide of choice, but ceased operations in 1968.)

In 1974 Schliessmann and Calheiros reported that *Ae. aegypti* had been eradicated from nineteen countries and territories comprising 73.5 percent of the initially infested land area of the hemisphere. Observing that such control of *Ae. aegypti* was clearly responsible for the absence of recent outbreaks of urban yellow fever in the Americas and citing recent outbreaks of dengue and jungle yellow fever, they echoed Soper's (1963) call for hemispheric eradication of *Ae. aegypti* and emphasized that in 1972 the directing council of PAHO had reaffirmed the goal of eradication. That reaffirmation was, in part, based on a cost-benefit analysis performed by the respected accounting firm of Arthur D. Little, Incorporated, under contract to PAHO, which concluded that for the entire hemisphere the benefits would exceed the cost, but that for some individual countries the cost would far exceed the benefits.

Evaluation

The spectacular successes of Gorgas, Cruz, and Soper in locally or regionally eliminating *Ae. aegypti* as a disease vector cannot be disputed. The reports of eradication of *Ae. aegypti* from seventeen countries comprising more than two-thirds of the originally infested land area of the Americas are indeed impressive. Nevertheless, Schliessmann and Calheiros (1974) were obliged to report that since 1965 there had been "reinfestations in 7 of the 17 countries that had previously achieved eradication of the vector. While the initial reinfestation of each country was limited in scope, difficulties have been experienced in combating the reappearance of the vector."

But was *Ae. aegypti* in fact eradicated from those countries? In light of the observed breeding of the mosquito in tree holes, rock depressions, leaf axils, and other natural water reservoirs, how much credibility can be attached to a claim of regional species eradication based largely upon surveys of man-made structures? The possibility that many of the "reinfestations" were in fact resurgences of residual local populations remains.

Believing that hemispheric eradication of *Ae. aegypti* was both desirable and practical, Schliessmann and Calheiros (1974) noted with concern that the elimination of yellow fever as a serious health problem in the Americas, coupled with the cost of achieving total hemispheric eradication, "has led to increasing skepticism by many of needs for continuing efforts to eradicate *Ae. aegypti.*" They also echoed the fear expressed by Reeves (1972) that the recent dengue epidemics in the Caribbean might "be an early warning of possible resurgence of epidemic urban yellow fever," particularly in Venezuela and the Caribbean. The latter concern resulted directly from the frequent resurgence of *Ae. aegypti* populations observed within a few years when control measures and surveillance were relaxed.

Fears of renewed epidemics of urban yellow fever cannot be taken lightly. The recent outbreaks of dengue in Cuba (Tonn, Figueredo, and Uribe (1982) illustrate the potential health problems posed by high *Ae. aegypti* populations. But is the appropriate solution *total species eradication?*

From 1977 to 1981, 878 cases of jungle yellow fever were reported in seven countries of the Americas (how many unreported cases must there have been?), yet not a single case of *Ae. aegypti*-transmitted yellow fever occurred even though "*Ae. aegypti* has not been eliminated and appears to be as successful as ever against control measures" (ibid). The last reported case of *Ae. aegypti*-transmitted yellow fever in the hemisphere occurred in Trinidad in 1954 (Gillette 1956), but even in that instance the evidence of *Ae. aegypti* involvement was hardly compelling.

Given the absence of serious yellow fever epidemics for several decades, neither human nature nor economic good sense can be expected to support an all-out *Ae. aegypti* eradication effort. But what of the Cuban dengue epidemics? Do they not argue for eradication? Let us examine a few of the details of that situation. At the beginning of the 1981 epidemic and prior to the imposition of emergency *Ae. aegypti* control efforts, the average countrywide *aegypti* premises index was 35 percent (Tonn, Figueredo, and Uribe 1982). After the first cases of dengue were reported in June 1981, aerial ULV application of malathion was carried out from June 6 to July 18. The "intensive attack phase" began on August 3. That phase included detailed inspection and source elimination and 100 percent pesticide coverage with temephos, fenthion, and malathion via 215 ULV ground fog generators, 3,961 portable mistblowers, and 4,407 hand spray pumps. By the end of August provincial *Ae. aegypti* premises indices ranged from 0.0 to 0.6 percent, and the last case of dengue was reported on October 10. A mop-up or consolidation phase featuring continued inspections and pesticide applications was begun in October 1981. The use of pesticides during various stages of suppression and the resulting premises indices are shown in table 16.1

Several observations concerning the Cuban experience that bear upon the concept of hemispheric eradication can be made: (1) Prior to the 1981 dengue epidemic, more than one-third of the premises in the country har-

Table 16.1 Pesticide Use and *Aedes aegypti* Prevalence during Dengue Vector Eradication Program in Cuba, 1981

| | Tons of Insecticide | | | Nationwide Average Premises Index at End of Phase (%) |
	Temephos (1% sand granules)	Fenthion (40% wdp)	Malathion (95% ULV)	
Beginning of dengue epidemic	—	—	—	35
Intensive attack phase	120.8	96.6	739.6	0.09
Consolidation phase				
Round 1	89.8	89.1	29.7	0.021
Round 2	90.7	0.0	29.2	0.012

Source: Tonn, Figueredo, and Uribe 1982

bored *Ae. aegypti*. Such a high index should never have been permitted and should not be allowed to occur again. But an all-out eradication effort is hardly required to maintain the index at a reasonable level—that is, below the critical index at which vector-mediated disease transmission can occur. (2) What is the critical index, and upon what factors does it depend? What studies have been undertaken to determine it? How can the costs of continual suppression be compared to the cost of one-time eradication if the critical index is not known with some accuracy? (3) A one-year effort involving physical source elimination and the countrywide application of about forty pounds of undiluted pesticide per square mile failed to eradicate *Ae. aegypti*. Can total eradication be achieved? (4) If an intense, pesticide-based eradication effort is maintained for several years only to be terminated when apparent, but not real, eradication has been achieved (the premises index may well be 0.0, but what of the tree-hole, beer can, or auto tire index?), how effective will pesticides be on a resurgent, high-density population of *Ae. aegypti* descended from the survivors of the war of eradication? (Recall the Puerto Rican experience reported by Fox and Bayona [1972] and Fox [1973]). (5) What is the long-term value of maintaining a premises index below a fraction of a percent (or even below 1 or 2 percent) if Cuba is surrounded by countries that have not eradicated *Ae. aegypti*?

The last question also applies to hemispheric eradication. In a world dependent upon international commerce, what is the long-term value of *Ae. aegypti* eradication from the Americas if the rest of the tropical and subtropical regions of the world are densely populated by the species? What is the real meaning of a zero *Ae. aegypti* index when there are several other genera of mosquitoes, most notably *Haemagogus* and *Sabethes* (de Rodaniche and Galindo 1957; Tonn, Figueredo, and Uribe 1982), that are vectors of the same diseases? These questions have been asked before. Sencer (1969), then director of the Centers for Disease Control in Atlanta, asked

Can the concept of hemispheric eradication [of *Ae. aegypti*] be supported in the absence of global eradication? Did not *Anopheles gambiae* become established in Brazil in days when transportation was much slower and much less frequent? From whence came *gambiae*? That same coast of Africa that led to the introduction of *A. gambiae* into Brazil is heavily infested with *Ae. aegypti* and the scene of the most recent epidemic of yellow fever (in Senegal in 1965). . . . But global eradication is not something that can be staged in one country and then another. It must be a simultaneous, total undertaking. Without simultaneous totality, reinfestations will be the rule rather than the exception. However, can it be proposed to the developing countries of the world that they place high priority on *Ae. aegypti* eradication in the absence of the immediate threat of disease? . . . A country struggling to provide even the barest of basic health services cannot be expected to indulge in the luxury of *Ae. aegypti* eradication for the benefit of other nations.

With such considerations, is hemispheric, or even national, species eradication a tenable and practical concept? Should intense efforts to achieve near zero urban and rural breeding indices be continually maintained at the risk of creating pesticide-resistant strains whose populations may explode and defy the usual suppression efforts? In the case of *Ae. aegypti*, would not a public health analogue of the agricultural pest manager's economic injury threshold be a more useful concept than species eradication? That human lives, rather than food and fiber, are the issue does not diminish the usefulness of the concept. Rather, it argues for more rigid and carefully conceived threshold values than those applied to agriculture. Vigilance and vaccination, not primary vector eradication, are the keys to the control of a disease that has an endemic, nonhuman reservoir and numerous secondary vectors.

The war against *Ae. aegypti* has been succinctly summarized by Tonn, Figueredo, and Uribe (1982). "Today, *Ae. aegypti* is still winning over the efforts of man despite new control equipment and insecticides as well as millions of dollars being spent annually against it." If this is so, should we heed the call for *Ae. aegypti* eradication when an effective and safe vaccine against yellow fever exists? Continual pesticide application is very expensive and requires hard currency. Development of resistance in the target pest continually increases that cost. Could not funds be more profitably invested in integrated pest management campaigns involving limited emergency use of insecticides, biological control, and public education as to the importance of breeding-site reduction and proper rubbish disposal?

Why is yellow fever absent from countries to the east of Africa even though *Ae. aegypti* is widely distributed and locally abundant (Christophers 1960)? Might there be an innocuous strain of *Ae. aegypti* that could be imported to competitively displace the disease-vectoring western strains?

These questions, many of which were posed years ago by Sencer (1969), directly challenge eradication and intense population suppression as the premier strategies for controlling diseases vectored by *Ae. aegypti*. It is time to consider whether less costly and more effective means are available. To champion eradication is to take refuge in the triumphs of the past and to ignore the realities of the present. Even in the unlikely event that one vector were to be eradicated, there are others.

The mosquito *aegypti* of the genus *Aedes*,
Was heard to remark ere it passed as a species,
You may dance with glee for eradicating me,
But what will you do with the genus *Sabethes*?

Literature Cited

Christophers, Sir S. R., 1960. Aedes aegypti *(L.), the yellow fever mosquito*. Cambridge: Cambridge University Press.

Cloudsley-Thompson, J. L. 1976. *Insects and history.* New York: St. Martins.

Eliason, D. A., J. W. Kilpatrick, and M. F. Babbitt. 1970. Evaluation of the effectiveness of ultra low volume application of insecticides against *Aedes aegypti* in Florida. *Mosquito News* 30:430–36.

Fox, I. 1973. Malathion resistance in *Aedes aegypti* from pressures on adults. *Mosquito News* 33:161–64.

Fox, I. 1977. Fenthion resistance in *Aedes aegypti* from selection pressure on larvae. *Mosquito News* 37:452–54.

Fox, I. 1980. Evaluation of ultra-low volume aerial and ground applications of malathion against natural populations of *Aedes aegypti* in Puerto Rico. *Mosquito News* 40:280–83.

Fox, I., and I. G. Bayona. 1972. Malathion resistant strains of *Aedes aegypti* in Puerto Rico in 1969. *Mosquito News* 32:157–60.

Gerberg, E. J., and W. M. Visser. 1978. Preliminary field trial for the biological control of *Aedes aegypti* by means of *Toxorhynchites brevipalpis*, a predatory mosquito larva. *Mosquito News* 38:197–200.

Gillette, H. P. S. 1956. Yellow fever in Trinidad: A brief review. *Mosquito News* 16:121–25.

Ignoffo, C. M., C. Garcia, M. J. Kroha, and T. Fukuda. 1980. Susceptibility of *Aedes aegypti* to four varieties of *Bacillus thuringiensis*. *Mosquito News* 40:290–91.

National Academy of Sciences. 1976. *Pest Control: An assessment of present and alternative technologies.* Volume 5. *Pest Control and Public Health.* Washington, D.C.: National Academy of Sciences.

Reeves, W. C. 1972. Recrudescence of arthropod-borne virus diseases in the Americas. *Pan Amer. Health Org. Scient. Pub.* 238:3–10.

Ribeiro, H. 1973. Entomological studies during the 1971 yellow fever epidemic of Luauda, Angola. *Mosquito News* 33:568–72.

de Rodaniche, E., and P. Galindo. 1957. Isolation of yellow fever virus from *Haemagogus mesodentatus, H. equinas* and *Sabethes chloropterus* captured in Guatemala in 1956. *Amer. J. Trop. Med. Hyg.* 6:232–37.

Schliessmann, D. J. 1967. *Aedes aegypti* eradication program of the United States: Progress Report 1965. *Amer. J. Publ. Health* 57:460–65.

Schliessmann, D. J., and L. B. Calheiros. 1974. A review of the status of yellow fever and *Aedes aegypti* eradication programs in the Americas. *Mosquito News* 34:1–9.

Sencer, D. J. 1969. Health protection in a shrinking world. *Amer. J. Trop. Med. Hyg.* 18:341–45.

Severo, D. P. 1956. Eradication of the *Aedes aegypti* mosquito from the Americas. *Mosquito News* 16:115–21.

Smith, H. H. 1951. Controlling yellow fever. In *Yellow Fever*. ed. G. K. Strode, 539–628. New York: McGraw-Hill.

Soper, F. L. 1935. Rural and jungle yellow fever: New public health problem in Colombia. *Rev. de Hig., Bogota* 4:49–84.

Soper, F. L. 1936. Jungle yellow fever, new epidemiological entity in South America. *Rev. de hyg. e saude publ., Rio de Janeiro* 10:107–44.

Soper, F. L. 1937. Geographical distribution of immunity to yellow fever in man in South America. *Amer. J. Trop. Med.* 17:457–511.

Soper, F. L. 1963. The elimination of urban yellow fever in the Americas through the eradication of *Aedes aegypti*. *Amer. J. Publ. Health* 53:7–16.

Soper, F. L., H. A. Penna, E. Cardoso, J. Serafim, Jr., M. Frobisher, and J. Pinheiro. 1933. Yellow fever without *Aedes aegypti:* Study of a rural epidemic in Valle de Chanaan, Espirito Santo, Brazil, 1932. *Am. J. Hyg.* 18:555–87.

Soper, F. L., and D. B. Wilson. 1942. Species eradication: A practical goal of species reduction in the control of mosquito-borne disease. *J. Nat. Malaria Soc.* 1:5–24.

Soper, F. L., D. B. Wilson, S. Lima, and W. S. Antunes. 1943. The organization of permanent nation-wide anti-*Aedes aegypti* measures in Brazil. New York: Rockefeller Foundation.

Strode, G. K., ed., 1951. *Yellow Fever.* New York: McGraw-Hill.

Taylor, R. M. 1951. Epidemiology. In *Yellow Fever,* ed. G. K. Strode, 427–538. New York: McGraw-Hill.

Tonn, R. J., R. Figueredo, and L. J. Uribe. 1982. *Aedes aegypti,* yellow fever and dengue in the Americas. *Mosquito News* 42:497–501.

Warren, A. J. 1951. Landmarks in the conquest of yellow fever. In *Yellow Fever,* ed. G. K. Strode, 1–37. New York: McGraw-Hill.

17

Malaria Mosquito in Brazil

James R. Davis and Richard Garcia

Anopheles gambiae Giles is an indigenous vector of malaria in tropical Africa. Its susceptibility to malaria infection and its close association with humans make it a vector of unparalleled importance. Despite modern scientific advances, *An. gambiae*-vectored malaria has resisted all attempts at eradication in Africa.

In 1930 *An. gambiae* was discovered in diked hay fields along the northeast coast of Brazil (fig. 17.1). Experienced malariologists immediately predicted dire consequences for Brazil and all of the new world tropics if *An. gambiae* became permanently established (Davis 1931; Shannon 1930). These fears were borne out when severe *An. gambiae*-vectored malaria epidemics occurred in Natal (1930, 1931) and the Assu, Apodi, and Jaguariba river valleys (1938, 1939) of northeast Brazil. The indigenous *Anopheles* vectors of malaria in Brazil had never caused such extreme morbidity and mortality. Entire agricultural valleys were devastated by the major epidemics of 1938 and 1939 (Soper and Wilson 1943). The tragic effects of *An. gambiae*'s presence in Brazil were countered by the rapid and unexpected eradication of the vector. By 1941 *An. gambiae* had disappeared from all of northeast Brazil, and malaria had returned to its original status before the ill-fated introduction of the vector (ibid).

Life History, Habits, and the Anopheles gambiae *Complex*

The biology of *An. gambiae* is similar to the biology of other *Anopheles* vectors. Adult females prefer to oviposit in relatively clean water with little or no current. The eggs are oviposited on the water surface and will hatch within two to six days, depending on temperature. Larvae feed on aquatic microorganisms and detritus floating on the water surface. Under normal tropical temperature conditions, larvae will complete four molts in about two weeks, and a pupal state of two to three days follows. Adult females do

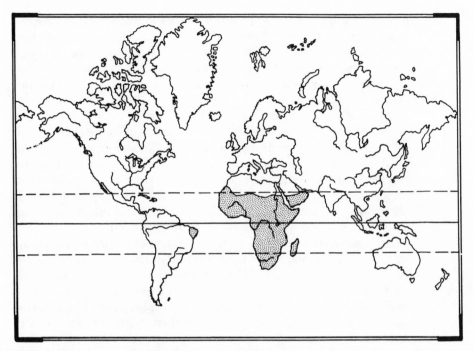

Figure 17.1 Native geographical distribution of *Anopheles gambiae* complex and zone of eradication in Brazil (adapted from Davidson et al. 1967).

not begin ovipositing until approximately seven days after emergence; thus, an entire generation from egg to egg requires about four weeks.

An. gambiae females are avid blood feeders and usually attack their vertebrate hosts in the evening or at night. A bloodmeal is usually required before each ovipositional egg development cycle. A single female may feed and oviposit many times during its life. Adult females become infected with the malarial parasites when taking a bloodmeal. The time between the uptake of the parasite by the mosquito and its development to the infective stage (extrinsic incubation period) usually takes ten to twelve days under tropical conditions. Most human malaria is transmitted by older females who have completed several ovipositional cycles.

When *An. gambiae* invaded northeast Brazil in 1930 it was thought to be a single homogenous species with only a few minor varietal differences. Since that time, taxonomists have recognized *An. gambiae* as a complex of six sibling species with slightly different biologies (Mattingly 1977). *An. gambiae A* (now known as *An. gambiae*) is the first member of the complex that was originally described by Giles in 1902. It can be distinguished from two of its five sibling species by morphological characteristics. But it can only be distinguished from the other three by chromosome banding, reproductive isolation, and behavioral differences. Thus, it is difficult if not impossible to determine which species invaded northeast Brazil. The minor biological and the occasional major behavioral differences between these closely related species can, however, aid speculation on which species was the actual invader.

An. gambiae A is a freshwater breeder found in the more humid areas of tropical Africa. It readily enters human habitations and is more easily controlled than the other species by insecticide applications to walls. *An. gambiae B* (now known as *Anopheles arabiensis* Patton) also breeds in freshwater and enters human habitations, but not nearly as frequently or persistently as *An. gambiae A*. Its range overlaps with that of *An. gambiae A* but also includes the drier steppes and savannahs of eastern and southern Africa. *An. gambiae C* (now known as *Anopheles quadriannulatus* Theobald 1911) is a relatively less common melanic species that is zoophilic and occurs on the high veld of Africa. Species D (as yet unnamed) is restricted primarily to the Rift Valley between Zaire and Uganda, where it is a local vector of malaria among Bambate pygmies (Service 1985). The two remaining members of the complex, *Anopheles melas* Theobald and *Anopheles merus* Don, are saltwater species; this fact readily separates them from the other species members.

Anopheles gambiae *in Northeast Brazil*

Northeast Brazil lies just south of the equator. Year-round temperatures are relatively mild, with minimums seldom falling below 20°C and maximums normally below 34°C. These temperatures are ideal for the invasion and establishment of the tropical malaria vector *An. gambiae*. Annual rainfall varies across the northeast, with about 400 mm/year in the Assu and Apodi river valleys, 800 mm/year in the Jaguariba river valley, and 1,000 mm/year in Natal along the eastern coast. The majority of the rainfall is concentrated in the wet season from February to August. Occasional sporadic rains occur from September to January and create temporary pools during the dry season. These alternating wet and dry seasons determined the available breeding sites for *An. gambiae*.

An. gambiae oviposited in nearly all types of freshwater during the wet season in northeast Brazil. During the dry season, however, the lush vegetation and flooded landscape became dry and parched, with water remaining in stream pools, coastal seepage areas, lakes, and artificial wells. At this time, *An. gambiae* larvae were found only in small, shallow freshwater ground pools exposed to the sun and free of aquatic vegetation. This microhabitat included shallow domestic wells, cattle hoofprints along lake or stream margins, irrigation seepage pools along the coast, roadside ditches, the isolated pools in streams and riverbeds, and a few natural pools at the edges of lakes. Thus, in the dry season the majority of the *An. gambiae* oviposition sites were created or modified by human agricultural and domestic activities and were located in the immediate vicinity of habitations.

In Africa, *An. gambiae* adults can be found resting in lush vegetation, under earth ledges, or in rock piles, where daytime temperatures are relatively low and humidities are high. But such resting sites do not exist in northeast Brazil during the dry season, when sparse vegetation and flat terrain predominate. The Brazilian seasonal variation in moisture affected *An. gambiae* adults. The glaring sun and wind associated with the dry season in northeast Brazil forced *An. gambiae* to seek shelter during the day. The most suitable microhabitats were found in human habitations.

The search for suitable daytime resting sites also led *An. gambiae* to enter motorized vehicles and boats. This behavior probably contributed to the major spread of *An. gambiae* to the Assu, Apodi, and Jaguariba river valleys. Earlier studies had shown that *An. gambiae* dispersal flight is usually limited to less than one-half miles (Evans 1927).

In northeast Brazil, the daytime resting habits of *An. gambiae* coincided perfectly with its nighttime feeding habits. *An. gambiae* was both endophilic and extremely anthropophagous, preferring humans above all other hosts. In fact, adult females were captured only inside human habitations (Soper and Wilson 1943).

The geography and topography of northeast Brazil had a major impact on the spread on *An. gambiae* and epidemic malaria. The isolated river valleys invaded by *An. gambiae* all drain directly into the Atlantic Ocean. These valleys are separated by low plateaus that were unsuitable for *An. gambiae* production (ibid). Such a plateau separates Natal, the original invasion site, from the rest of central and southern Brazil. To the northwest the valleys are connected by a moist, narrow coastal plain that may have served as a natural route for the spread of *An. gambiae*. Only a small portion of northeast Brazil was suitable for *An. gambiae* breeding and adult survival.

In light of this biological information, the slow spread of malaria throughout the northeast, and the severity of the epidemics, it seems unlikely that *An. gambiae A.* was the invading species.

The barren low plateaus and dry season climate of northeast Brazil would have provided less than optimal conditions for the survival of species *A*. These factors would have forced it to remain in close association with human settlements, where shelter and water sources were available. For a species that is both endophilic and anthrophilic, such conditions would have resulted in serious local epidemics of malaria—which did occur. Furthermore, because *An. gambiae A* would not have been able to establish populations away from human settlements in such a dry climate as northeast Brazil, malaria would have spread slowly in the region—a pattern that was borne out.

Malaria epidemics would probably have been less pronounced if the less anthrophilic species B (*An. arabiensis*) had invaded northeast Brazil. This species is adapted to drier climates, and the conditions in northeast Brazil were more appropriate for its survival. It would have been able to become established away from human settlements, and a more rapid geographic spread of malaria would have ensued. It is likely that eradication would also have been more difficult.

The zoophilic species C, *An. quadriannulatus*, would not have been a candidate for transmitting malaria to humans, whereas species D would have been a very unlikely immigrant to Brazil because of its restricted range in the Rift Valley in Africa. The remaining two members of the complex, *An. melas* and *An. merus*, are not possibilities because neither breeds in fresh water.

Control

Larval sampling, larval control, adult sampling, and adult control were all performed by separate work crews. Each crew was responsible for a precisely delimited geographic area and a clearly defined job. The limited number of dry season oviposition sites and the strict endophilic behavior of *An. gambiae* adults simplified both sampling and control methods; "*gambiae* breeding during the dry season is so definitely limited to certain types of water collections that failure to find larvae in these collections may be regarded as good evidence of its absence from the region. Likewise, *gambiae* is so highly domestic in northeast Brazil that the failure to capture adults in houses in also good presumptive evidence that it is not present in the community" (ibid).

Both larval and adult samples were used to determine the extent of spread, the need for and efficiency of local control measures, and the success of regional eradication efforts. Larvae were collected by standardized dipping and sent immediately to field laboratories for identification. As insurance against mistakes, all samples were reviewed by a control laboratory.

Adults were sampled in human habitations by three methods. Initially, hand sampling with suction bulbs was tried, but adults were difficult to see on the dark rough walls. Later, the use of inverted umbrellas and pyrethrum knockdown sprays was discovered to be more effective for capturing adults. A third method dispensed with the umbrellas. After all doors and windows were closed, each room was sprayed systematically, and the sprayed adults were collected directly off the floor. Large numbers of *An. gambiae* were occasionally captured by these pyrethrum knockdown methods, and the majority were frequently found in the sleeping quarters. According to Soper and Wilson (1943), "A rapid four minute application of spray with umbrella capture produced ninety-one *gambiae*. An additional sixty-seven *gambiae* were recovered from some four meters of floor surface where they had missed the umbrella." Indoor adults samples were found to be a more sensitive indicator of the presence of *An. gambiae* than were careful searches of larval breeding sites.

Serious efforts to eradicate *An. gambiae* did not begin until 1939, after the major epidemics in the Assu, Apodi, and Jaguariba river valleys. Prior to this, larval control measures emphasized habitat modifications via drainage, filling, or flooding of potential oviposition sites. Occasionally, insecticidal oils were used where these methods proved too costly. When Fred Soper assumed command of the eradication project in 1939, "the attack on *An. gambiae* was frontal and direct" (ibid). "Naturalistic" methods were replaced by Paris green for larval control and pyrethrum for adults. Soper believed that insecticides were the easiest, cheapest, and fastest control method for the available labor force (ibid).

Paris green was used in wet, moist, or dry formulations, depending upon the season and the availability of water. Each formulation was mixed by hand using local sources of water or road dust. Knapsack pressure sprayers or hand casting were used to distribute the insecticide. All potential breeding sites (including domestic and agricultural water supplies) were treated on a weekly or biweekly schedule until local eradication had been achieved.

Control of *An. gambiae* adults was similar to adult sampling methods. Since residual DDT was not yet available, the walls, ceilings, and floors of all man-made shelters were sprayed on a weekly basis with pyrethrum in diesel oil. Disinsectization of motor vehicles, trains, airplanes, and boats also relied on pyrethrum space sprays. All vehicles leaving the infested zones were sprayed with an excess of insecticide and then closed for five minutes. Thirty-three disinsectization posts were established on the periphery of the infected zones.

History of Eradication Efforts

An. gambiae is presumed to have crossed the Atlantic in a French destroyer that was making a fast mail run from Dakar in Africa to Natal in northeast Brazil (ibid). Shortly after *An. gambiae* was discovered in March 1930, there was an explosive outbreak of malaria in Natal. By the following January a second, more serious, and widespread epidemic had begun. Ten thousand cases were reported from a population of twelve thousand in a six square-kilometer area.

By March 1931, control measures were begun by the Yellow Fever Service of Brazil. At the time, the Yellow Fever Service was engaged in an eradication battle with *Aedes aegypti* L., the major vector of yellow fever, and could not afford to launch a major attack against *An. gambiae*. Nonetheless, by April 1932, the limited control effort of the service had successfully eradicated *An. gambiae* and malaria from the immediate vicinity of Natal. This initial success was extremely important, because Natal was a transportation center to the rest of Brazil. Although *An. gambiae* had been eradicated from Natal by mid-1932, however, it had by this time spread into rural areas to the north and west. During 1932 investigators located *An. gambiae* populations by pursuing rumors of severe malaria outbreaks in these local regions. But the eradication effort was not continued in rural areas for several reasons. The resources and attention of the Yellow Fever Service were focused on *Ae. aegypti* and the devastating urban epidemics of yellow fever in central Brazil. After malaria was eradicated from the populated and politically astute state capital of Natal, the pressure on the Yellow Fever Service to do the same for rural areas was reduced. Severe droughts in northeast Brazil in 1932 further distracted attention from the serious but limited epidemics occurring in the rural regions.

From 1932 to 1937 *An. gambiae* was discovered infrequently and apparently caused no major outbreaks of malaria. During this period *An. gambiae* was slowly expanding its range into more favorable agricultural valleys to the northwest of Natal. By 1938 severe malaria epidemics occurred in the rural and village populations in the Assu, Apodi, and Jaguariba river valleys. Soper and Wilson (1943) quote Federal Health Delegate Valerio Konder: "The entire population is very poor, living from hand to mouth on the day's wages; on the day a man does not work, he and his family do not eat. This is one of the principal causes of the high mortality of this outbreak of malaria, many of the deaths being due to starvation as well as to lack of treatment. . . . It would take too long to narrate here all the sad details of the trip through this stricken region where it took will power to continue the program of investigation instead of giving way to the insistence of the ragged and hungry multitude which pleads for mercy at the side of the road, trying to block our passage and induce us to examine the sick who were dying in nearby houses."

By May 1938, local governments appealed to the central Brazilian government for assistance. The Antimalaria Service was organized in August 1938 with a budget of $50,000. The new agency had to set up laboratories, train personnel, and develop eradication strategies while the local population was suffering severely from the malaria epidemic. A major portion of the budget was absorbed by drug purchases, and numerous difficulties arose in procuring necessary equipment. Both larval and adult insecticides were not available in large quantities, so the eradication effort was initially limited to environmental modifications and insecticidal oils. Little, if any, success was being achieved against *An. gambiae*.

By 1939 the International Health Division of the Rockefeller Foundation expressed interest in *An. gambiae*. The Rockefeller Foundation was already involved in the project to eradicate *Ae. aegypti* and yellow fever. After negotiations, the Brazilian government agreed to a cooperative endeavor. Under the agreement, the new Malaria Service of the Northeast received $250,000 from the Brazilian government and free access to all equipment and personnel of the Yellow Fever Service. The Rockefeller Foundation provided $100,000, and administrative authority was given solely to Fred Soper of the foundation. Soper (1949) mentioned these negotiations in a later publication: "It is a pleasure to record here the attitude of the government representatives in these discussions. Both agreed to recommend that the program for the eradication of *gambiae* be undertaken as a cooperative project, even though this would take from them direct administrative control." It was extremely unusual for Brazilian government officials to allow the Rockefeller Foundation to assume full administrative authority over a major project in their own country. The terms of the contract (Soper and Wilson 1943) reveal the authority of the Rockefeller Foundation in the developing countries of that era. This institutional arrangement provided the necessary expertise and resources for eradicating *An. gambiae* while managing to avoid the unnecessary delays often associated with many large government programs.

The aggressive approach used to initiate the Malaria Service of the Northeast was continued throughout the eradication effort. A hierarchical and regimented organization was designed by Soper to reflect his belief that the major problem in vector eradication in insuring that field personnel use pesticides effectively. To achieve effective pesticide use Soper (1949) felt that: (1) the infested area must be divided into well-defined zones; (2) responsibility must be clearly assigned in well-defined jobs; (3) field personnel must maintain written records of all activities; (4) all work must be checked frequently by supervisors; (5) larval and adult sampling should be done by separate crews as a double check; (6) the simplest control methods should be chosen; and (7) all supervisory instructions should be in written form. The structure of the Malaria Service of the Northeast has been described by Soper and Wilson (1943).

By the end of 1940, *An. gambiae* was eradicated from northeast Brazil. This incredible and unexpected eradication was completed in fewer than two years. Malaria returned to its normal endemic status with only an occasional minor epidemic caused by the indigenous vectors that had survived the eradication effort.

Evaluation

Although the eradication of *An. gambiae* was extremely successful, it was not without faults. The field personnel of the Malaria Service of the North-

east and the general public did not completely escape the hazards of pro-
longed and excessive pesticide exposure. Arsenic poisoning was common
among men in larvicide crews. Attempts were made to train personnel in the
safe and effective use of Paris green, but these consisted of little more than
instructions to avoid wind-blown dust, to take frequent baths, and to maintain clean fingernails. During the dry season dust formulations required on-
site hand mixing, and nearly all water was being treated regularly with Paris
green. Workers who reacted severely and persistently to Paris green were
transferred from larvicide crews to some other branch of the service.

The philosophical approach to the general public was similar. Soper and
Wilson (1943) write: "Fortunately popular opinion did not attribute any
human ailments to Paris green and only six accidents were reported during
the application of some 260 tons in northeast Brazil." Two people used Paris
green to commit suicide, and three children were poisoned, one of whom
did not recover. In areas where *An. gambiae* had shown its power as a vector,
there was relatively little resistance to the application of eradication measures. *An. gambiae*-transmitted malaria was an effective schoolmaster; only
in the frontier areas where *gambiae* had not yet penetrated was there resistance to the use of Paris green, and this was because of livestock and
poultry deaths. Some of these deaths were later attributed to undiagnosed
cases of brucellosis. Farmers who refused to cooperate with larviciding
crews were persuaded by more direct means. For example, insecticidal oils
that make domestic and agricultural water sources unfit for use were used as
larvicides instead of Paris green as a disciplinary measure to force compliance with the protection of water holes against infestation (ibid).

The successful eradication of *An. gambiae* from northeast Brazil can be
attributed to administrative efficiency, as well as to certain biological characteristics of the vector. *An. gambiae A* was an introduced species that was not
fully adapted to northeast Brazil in either its larval or adult stages. Its
potential habitats were severely limited during the dry season, and it was
unable to spread rapidly into more favorable areas. Eight years after the
invasion, the distribution of *An. gambiae* in northeast Brazil was still limited to a narrow geographic area.

An. gambiae also failed to become established in inconspicuous natural
breeding sources. Its peridomestic oviposition, resting, and feeding habits
were important both to its success as a malaria vector and to its vulnerability
to eradication. *An. gambiae* was easy to sample in the dry season; control
efforts could be accurately directed, and local eradication verified.
Pyrethrum was an extremely effective insecticide for *An. gambiae* adults.
Control was simplified by the limited number of dry season oviposition
sites and the strictly endophilic behavior of adults. Had *An. gambiae B*
invaded northeast Brazil, the eradication might not have proceeded nearly
as effectively.

The most important characteristic of the *An. gambiae* eradication program was the extreme regimentation of resources and personnel under the direct authority and supervision of Soper. The possibility that *An. gambiae* would spread throughout the new world tropics had prompted the Rockefeller Foundation to intervene and then to persuade the Brazilian government to relinquish control over the eradication project. The trained personnel and equipment available through the Yellow Fever Service assured the rapid and efficient initiation of control activities. Soper's regimented organization insured that field crews used insecticides effectively and thoroughly. The specter of severe epidemic malaria allowed and justified such a concentration of power and a disregard for the environmental and public health impacts of excessive insecticide use. Without the exercise of such power, the eradication of *An. gambiae* in northeast Brazil would probably have failed.

Literature Cited

Davidson, G., H. E. Paterson, M. Coluzzi, G. F. Mason, and D. W. Micks. 1967. The *Anopheles gambiae* complex. In *Genetics of Insect Vectors of Disease,* ed. J. W. Wright and R. Pal, 211–50. New York: Elsevier.

Davis, N. C. 1931. A note on the malaria carrying anophelines in Belem, Para, and in Natal, Rio do Norte. *Rivista di Malariologia* 10:43–51.

Evans, A. M. 1927. A short illustrated guide to the anophelines of tropical and south Africa. London: University Press of Liverpool.

Mattingly, P. F. 1977. Names for the *Anopheles gambiae* complex. *Mosquito Systematics* 9:323–28.

Service, M. W. 1985. *Anopheles gambiae:* Africa's principal malaria vector, 1902–1984. *Bull. Entomol. Soc. America* 31(3):8–12.

Shannon, R. C. 1930. O apparecimento de una especie africana de Anopheles no Brasil. *Brasil-medico* 44:515–16.

Soper, F. L. 1949. Species sanitation and species eradication for the control of mosquito borne diseases. In *Malariology,* ed. M. F. Boyd. Philadelphia and London: W. B. Saunders.

Soper, F. L., and D. B. Wilson. 1943. *Anopheles gambiae* in Brazil, 1930 to 1940. New York: Rockefeller Foundation.

Contributors

Daniel J. Clair
Texas Department of Agriculture
Austin, Texas

Kent M. Daane
Division of Biological Control
University of California
Berkeley, California

Donald L. Dahlsten
Division of Biological Control
University of California
Berkeley, California

Nita A. Davidson
Department of Entomology
University of California
Davis, California

James R. Davis
Division of Biological Control
University of California
Berkeley, California

Steve H. Dreistadt
Division of Biological Control
University of California
Berkeley, California

Richard Garcia
Division of Biological Control
University of California
Berkeley, California

Francis Geraud
Universidad del Zulia
Maracaibo, Venezuela

Charles W. Getz, IV
Deputy Attorney General
California Department of Justice
San Francisco, California

J. Kenneth Grace
Faculty of Forestry
University of Toronto
Toronto, Ontario, Canada

Kim A. Hoelmer
Division of Biological Control
University of California
Berkeley, California

Vicki L. Kramer
Division of Biological Control
University of California
Berkeley, California

Olaf Leifson
California Department of Food
and Agriculture (Retired)
Sacramento, California

E. Phillip LeVeen
Consultant and Lecturer,
Department of Conservation
and Resources Studies
University of California
Berkeley, California

Hilary Lorraine
USDA, APHIS, PPQ
U.S. Embassy, Guatemala
APO Miami, Florida

Jill W. Lownsbery
Forest Pest Management
USDA, Forest Service
Flagstaff, Arizona

Stephen A. Manweiler
Division of Biological Control
University of California
Berkeley, California

Laura D. Merrill
Department of Entomology
University of California
Berkeley, California

Donald R. Owen
California Department of Forestry
Redding, California

John H. Perkins
The Evergreen State College
Olympia, Washington

Michael J. Pitcairn
Department of Entomology
University of California
Davis, California

Jerry Scribner
Attorney-at-Law
Sacramento, California

Durward D. Skiles
Consultant
Martinez, California

Nick D. Stone
Department of Entomology
Texas A&M
College Station, Texas

Donald C. Weber
Department of Entomology
University of Massachusetts
Amherst, Massachusetts

Larry R. Wilhoit
Department of Entomology
University of California
Berkeley, California

John S. Yaninek
International Institute for Tropical
Agriculture
Ibadan, Nigeria

Index

ERADICATION OF EXOTIC PESTS

ANALYSIS WITH

CASE HISTORIES

DONALD L. DAHLSTEN

RICHARD GARCIA

EDITORS

HILARY LORRAINE

ASSOCIATE EDITOR

Exotic pests frequently create problems for commercial agriculture, forestry, and health in both the developed countries and the Third World. Two common ways of dealing with these pests are eradication and biological control, and a great deal of controversy has centered on which of these methods to use. In this book, experts in pest control discuss the most recent methods for and implications of eradicating these pests.

The book begins with essays dealing with broad questions involving scientific, social, economic, institutional, and public policy issues surrounding eradication programs. The authors then present case histories of twelve eradication programs, ranging from yellow fever and malaria mosquitoes, which have long been recognized as public health problems, to potential urban, agricultural, and forest insect pests such as the Japanese beetle, gypsy moth, and imported fire ants. Two plant disease pro-